D1067864

The Fifth Branch

THE FIFTH BRANCH

Science Advisers as Policymakers

Sheila Jasanoff

Harvard University Press
Cambridge, Massachusetts
London, England
1990

Printed in the United States of America
10 9 8 7 6 5 4 3 2 1

This book is printed on acid-free paper, and its binding materials have been chosen for strength and durability.

Library of Congress Cataloging-in-Publication Data

Jasanoff, Sheila.
The fifth branch : science advisers as policymakers / Sheila Jasanoff.
p. cm.
Includes bibliographical references.
ISBN 0-674-30061-0 (alk. paper)
1. Science and state. 2. Technology and state. 3. Technocracy.
I. Title.
Q125.J33 1990 90-32942
338.9'26—dc20 CIP

Preface

The bomb that dropped on Hiroshima on August 6, 1945, shattered not just bodies and buildings but also the myth that scientists can remain detached from the uses of their knowledge. History has remembered, even if it has not invariably honored, the eminent physicists who advised Secretary of War Henry L. Stimson on possible targets for the bomb. In the years since those fateful events, scientific advice-giving has become at once a more prosaic and a more pervasive process, rarely examined or exposed to public view, yet exercising growing influence on our daily lives. Should we eat supermarket apples, use hairspray, drive cars in inner cities, incinerate our wastes, generate nuclear energy, release genetically engineered organisms into the environment? We expect our political decisionmakers to seek expert advice on all such matters, but few of us any longer know who the advisers are or how much they influence public policy.

This book was written in an effort to bring the modern scientific advisory process out of the shadows and into the limelight of public policy analysis, where it rightfully belongs. My inquiry into the role of such committees began in 1985, at a time when the scientific credibility of the agencies was under attack from both the right and the left of the political spectrum. There was general agreement that regulators had fallen short in their efforts to obtain high-quality science pertaining to health, safety, and environmental policy. More troubling still were charges that federal agencies were purposefully intermingling politics with science, and a consensus had begun to emerge that the independent scientific community should play a more vigorous role in holding agency officials to higher standards of scientific accountability. Critics urged, in particular, that the scientific and technical basis for rulemaking should be subjected to much more systematic peer review in order to prevent politically motivated distortions of science.

Ironically, however, the credibility of science has also suffered unexpected damage in recent years. Lapses in the processes of scientific self-regulation, signaled by revelations of fraud and misconduct at some of our premier research institutions, have called into question the capacity of scientists to vouch for the integrity of their published findings. Problems of quality assurance and quality control in science are recognized as extending well beyond the confines of the regulatory process. As a result, the idea that agencies can cure their scientific deficiencies simply through greater reliance on peer review and expert consultation has come to be perceived as unduly optimistic, even naive.

This book, then, was conceived and written in response to a crisis of confidence surrounding not only political but scientific authority. The questions it addresses thus have more than common urgency. How can decisionmakers make sure that policies affecting the health and safety of the American public and the quality of the national and global environment are not perceived as politically self-serving and scientifically unsound? How should agencies minimize the potential for bias and conflicts of interests in their dealings with advisory committees? Can scientists be more deeply involved in the regulatory process without risking their political neutrality or, worse yet, actually making policy, and thereby eroding democratic controls on decisionmaking? Alternatively, can the scientific advisory process be organized in ways that further public participation but do not lead to the capture of science by political interests?

Questions such as these could not be answered without empirical research on the performance of advisory committees in the regulatory process. Much of the book is therefore devoted to an analysis of policy decisions in which scientific advice played an important role. Not surprisingly, the cases that proved to be most interesting were also the most protracted and ambiguous. The majority of the cases I consider preoccupied the regulatory establishment for a number of years, requiring, in most instances, repeated and often inconclusive rounds of consultation with expert committees before a final decision was reached. To understand the role of science in these cases, it became necessary to look back—often far back—not only into the regulatory agency's patterns of seeking scientific advice but also into its past successes and failures and its relations with Congress, the White House, other agencies, the public, and the courts.

My intention was not to create a mere compilation of fascinating regulatory case histories or to write a practical handbook for policymakers. A theoretical concern that guided the book was the need for a richer conceptual framework, and even a more differentiated vocabulary, for discussing the dilemmas confronted by scientific advisory committees. Concepts furnished by the literature on law and public policy—such as the problematic term "science policy"—were inadequate for this purpose. In recent years, there have been notable advances in our understanding of the social processes by which scientific knowledge is produced and certified as legitimate. We have become aware of the socially constructed nature of scientific reality and of the intermingling of facts and values in disputes arising at the frontiers of science. Such insights have not figured prominently in the policy analysis literature, though they are clearly relevant to an exploration of the use of science by policymakers. In this book I draw liberally on findings from the social studies of science, and I hope in the process to open the door to further interdisciplinary research linking the theoretical world of science studies with the practical concerns of science policy.

At the heart of the book is a critique of two commonly accepted paradigms for controlling the use of science by regulatory agencies: the "technocratic" approach, which looks to scientists as primary validators of policies with high technical content, and the "democratic" approach, which views broad public participation as the antidote to abuses of expert authority. Neither approach, in my view, takes adequate account of the nature of science or of politics. Yet players in the regulation game, aware that decisionmaking can be coopted through strategic choices of procedural and institutional design, have frequently championed one or the other model in pursuit of their immediate political objectives. This book argues for a richer and more realistic account of the role of expertise in public decisions. My aim has been to locate the discussion of advisory committees, and recommendations for their use, within a solid framework of theoretical and empirical scholarship on the relations between science and regulation.

Most of my empirical research centered on just two federal regulatory agencies, the Environmental Protection Agency (EPA) and the Food and Drug Administration (FDA), and their advisory committees. In assessing the role of these bodies, I had to look in sometimes microscopic detail at the dynamics of particular committee meetings, the interpretation of particular studies, and the impact of particular scientific claims on policy. But the close focus on two agencies and their technical advisers should not distract attention from the book's deeper purpose, which is to illuminate and critique a part of the process by which our society makes choices about science and technology. Controversies involving expert advisory committees serve, in the end, as a lens for looking into the limits of participatory decisionmaking in an age of growing technological complexity. Is it possible—and, if so, how—to preserve a meaningful role for the lay public and its political representatives in areas of decisionmaking that seem increasingly less accessible to those who do not possess specialized knowledge? This is an issue that demands wide public debate and serious engagement. Otherwise, an ever larger sphere of public policy is in danger of escaping effective criticism and popular control.

Like most books, this one could not have been written without help and cooperation from many quarters, and it is a pleasure to acknowledge these obligations publicly. Research for the study was generously supported by a grant from the National Science Foundation's Division of Policy Research and Analysis. The Program on Science, Technology and Society at Cornell University provided invaluable administrative support throughout the project. Sandra Kisner applied her considerable talents to preparing the early drafts of the manuscript, and Rosanne Mayer, Nancy Dickson, Tana Linn, Olive Lee, and Robert Speel extended by much more than fivefold my own ability to collect and synthesize the research materials needed for a study of this scope. I am also deeply indebted to the Centre for Socio-Legal Studies at Oxford University for providing a hospitable and peaceful haven during the semester when I began writing the book.

Many individuals responded to my inquiries with generous contributions of information, insights, and time. I am particularly grateful to more than thirty scientists, agency officials, industry and public interest representatives, and academics whom I interviewed in the course of the project. Without the benefit of their expertise and specialized knowledge, this book could never have been written, and I regret that it is not possible to thank them all by name. Some debts, however, demand a more personal acknowledgment. I would like to thank, in particular, Edward Burger, J. Clarence Davies, Richard Merrill, and Grover Wrenn for their invaluable guidance during the initial planning phase. Their suggestions concerning the choice of cases and advisory mechanisms were especially helpful. My deepest gratitude is reserved for those demanding but friendly critics who took time and trouble to read and comment on substantial portions of the manuscript: Richard Andrews, Edward Burger, J. Clarence Davies, John Graham, Harry Marks, Richard Merrill, M. Granger Morgan, Christopher Wilkinson, Terry Yosie, and, most of all, Thomas Gieryn and Brian Wynne. Although I have not always followed their advice as closely as they might have wished, their informed criticism inspired many refinements and improvements in the text. Any remaining errors and infelicities are, of course, my sole responsibility.

My editors at Harvard, Howard Boyer and Camille Smith, not only provided many helpful suggestions but briskly treated all symptoms of authorial paralysis that threatened to delay the publication of the book.

The final revisions on this manuscript were completed in the shadow of my father's unexpected death in the summer of 1989. A prolific writer to the end of his life, he took great satisfaction in his daughter's career with words, and the appearance of this book would have given him much pleasure. To his memory and to my mother I dedicate the finished work.

Contents

The Fifth Branch

1

Rationalizing Politics

Scientific advisory committees occupy a curiously sheltered position in the landscape of American regulatory politics. In an era of bitter ideological confrontations, their role in policymaking has gone largely unobserved and unchallenged. Advisory committees are generally perceived as an indispensable aid to policymakers across a wide range of technical decisions. They offer a flexible, low-cost means for government officials to consult with knowledgeable and up-to-date practitioners in relevant scientific and technical fields, supplementing the unspecialized and sometimes pedestrian expertise available within the executive branch. Perhaps most important, they inject a much-needed strain of competence and critical intelligence into a regulatory system that otherwise seems all too vulnerable to the demands of politics. It is hardly surprising, then, that in most programs of health, safety, and environmental regulation, consultation between agencies and advisory committees has become almost routine, even when not required by law. The proposition that science-based decisions should be reviewed by independent experts strikes us today as hardly more controversial than the proposition that there is no completely risk-free technology.

Yet, given the centrality of their role in the regulatory process, the activities of scientific advisers are poorly documented and their impact on policy decisions is difficult to understand or evaluate. If a cardinal function of advisory committees is to take the politics out of policymaking, then a survey of the American regulatory scene for the past twenty years casts doubt on their efficacy. Not only were regulatory decisions during this period particularly prone to legal and political challenge, but a remarkably high percentage of the challengers targeted the quality and sufficiency of the agencies' technical arguments. Evidence of consultation with expert committees rarely proved sufficient to silence controversy. These are some of the paradoxes that I set out to examine in this book. Why does a regulatory process so strongly committed to rational decisionmaking and use of expert knowledge so frequently fail to produce consensus over the use of science? What are the factors that inhibit scientific advisory committees, in particular, from containing or closing technical disputes? Conversely, is it possible to identify conditions under which an advisory committee's intervention will be accepted as authoritative by other players with a stake in policymaking?

In the U.S. regulatory process, support for advisory mechanisms coexists with widespread disagreements about how to select advisers, how to frame issues for their consideration, and how much weight to give to their recommendations. The absence of such agreement points to still more basic differences about allocating scientific and technical power between experts and the lay public, among competing political interest groups, and between citizens and the state. An investigation of the politics of scientific advice thus provides an avenue for exploring some of the enduring conflicts between democratic and technocratic values in this country's public and political life.

To establish the historical and analytical framework for the remainder of the book, I outline in this introductory chapter some factors that constrain the performance of modern scientific advisory committees as legitimators of public policy. I begin by describing the institutional and political environment within which advisory committees carry out their business. The recent explosive growth of scientific advising has taken place against a backdrop of growing public concern about technological hazards, accompanied by diminished trust in government and ambivalence about the place of experts in political decisionmaking. The second part of the chapter relates these problems to recent scholarly findings about the admixture of science and values in regulatory proceedings and about the contingent and negotiated character of scientific knowledge.

The Rise of Social Regulation

The changing complexion of governmental regulation in the 1970s provides part of the context in which scientific advisory committees operate. The rapid expansion of social regulation in this period created a host of new agencies and expanded the reach of federal regulatory activity across a much wider cross-section of commerce and industry.[1] Simultaneously, the nature of technical decisionmaking in the agencies underwent profound changes. To protect the public against hazardous and environmentally harmful technologies, fledgling agencies were asked to undertake ever more complex predictive analyses of the risks and benefits of regulation. The costs of controlling risk grew in a seemingly inverse relationship to the certainty of harm. Decisions of unprecedented socioeconomic impact seemed increasingly to be based on imperfect knowledge developed by inexperienced administrators through novel and untested scientific techniques. Beginning in 1981, moreover, the Reagan administration identified government regulation as the prime impediment to technological innovation and as an important contributor to America's flagging performance in the world economy. Under these combined pressures, public faith in the professionalism and specialized expertise of regulatory agencies, and in the legitimacy of their decisions, gradually eroded.[2]

The new programs of social regulation required most policy decisions to be founded on an explicit trade-off between risks to health or the environment and the economic and social costs of regulation. Since neither side of the calculation could be precisely estimated, suspicion grew that regulators were arbitrarily overstating one or the other in order to reach predetermined results. Under these circumstances, it became difficult for agency officials—seen by many as an overly powerful fourth branch of the government—to avoid creating the impression that they were manipulating scientific knowledge and shielding fundamentally political choices behind the pronouncements of a still more inscrutable "fifth branch" of technical experts.

The perception that regulators were permitting political considerations to corrupt the integrity of their scientific analyses spread across the entire political spectrum. It was a common complaint of industry and members of the scientific community during the 1970s that the Environmental Protection Agency (EPA) and the Occupational Safety and Health Administration (OSHA) had systematically distorted their assessments of cancer risk so as to build the case for more regulation. Others charged regulators with selectively using expert knowledge to enlarge their own political agendas. According to one account, for instance, the Food and Drug Administration (FDA) deliberately used inadequately validated animal studies of saccharin to propose that the artificial sweetener be banned.[3] By contrast, in the early years of the Reagan administration, environmentalists accused EPA of introducing a pro-industry and antiregulatory bias into its principles for assessing carcinogenic risk.

The scientific credibility of the regulatory agencies was also hurt by occasional well-publicized discoveries of research fraud and misrepresentation of data. Although such cases were not numerous, some of them were flagrant enough to attract attention from Congress and the media. Beginning in 1976, the discovery that Industrial Bio-Test Laboratories (IBT), a private testing concern, had systematically falsified data on hundreds of toxicology tests sent shock waves through the regulatory establishment.[4] A five-year investigation of IBT led to criminal prosecutions against three of the company's top executives,[5] and, more seriously from the standpoint of public health, to determinations in the United States, Canada, and Sweden that numerous insecticides and herbicides registered on the basis of IBT data should be withdrawn from commercial use.

In other instances, regulatory agencies were directly implicated in charges of misconduct or incompetence in research related to public policy. EPA, for example, was seriously embarrassed by allegations of fraud in the health studies it commissioned at Love Canal. A similar problem occurred when EPA learned that the principal studies underlying a proposed air quality standard for carbon monoxide had been conducted by a scientist who was suspected of fabricating data and had been debarred from doing research for two other agencies. Allegations of scientific impropriety also helped to discredit the standard for benzene promulgated by OSHA in 1980 and were a factor in the judicial reversal of that action. Episodes such as these damaged the already fragile credibility of the agencies as scientific and technical decisionmakers.

Science and Policymaking

The role that science and scientists have played in policymaking during the past few decades provides another indispensable piece of the context for a study of contemporary scientific advisory committees. What do we actually know about the uses of science in policy decisions, and what does this suggest about the place of experts in the regulatory process?

Writing about scientific advisers back in 1964, Harvey Brooks almost casually introduced into the literature a conceptual distinction that has since served to anchor most discussions of the relationship between science and government. The functions of advisers, Brooks suggested, can be loosely divided under the headings of "science in policy" and "policy for science":

> The first is concerned with matters that are basically political or administrative but are significantly dependent on technical factors—such as the nuclear test ban, disarmament policy, or the use of science in international relations. The second is concerned with the development of policies for the management and support of the national scientific enterprise and with the selection and evaluation of substantive scientific programs.[6]

The distinction, as Brooks himself acknowledged, does not necessarily correspond to a clear difference. Policies *for* science need input from respected scientists if they are to be regarded as credible. Similarly, social policies *in* which science plays a substantial role—clean air standards, licenses for nuclear power plants, pesticide registration, permits for genetic engineering experiments—invariably exercise a secondary impact on the nation's policies for managing its scientific and technological resources. Nonetheless, the two categories serve as useful headings for organizing the literature on scientific advising and public policy.

The need for better advisory mechanisms has frequently been noted in works dealing with policymaking for science, but such studies have tended to downplay the controversies and conflicts that arise from attempts to use scientific information in policy decisions. Characteristic of this genre is William T. Golden's massive compilation of essays on science advice to the federal government.[7] The book persuasively argued for a restoration of the President's Science Advisory Committee (PSAC) so as to ensure more systematic technical input to the nation's policies for science, especially with respect to expenditures for military and civilian research. But of the sixty-seven contributors who addressed the needs of the president and the executive branch, including several former PSAC members, not one was centrally concerned with the delivery of scientific and technical information to federal regulatory agencies.[8] Preoccupied with the problem of too little scientific advice, particularly in the sphere of presidential decisionmaking, Golden's book skirted or overlooked the complexities that arise when contested science is factored into policy.

Much more closely related to the concerns of this book is a small but growing body of work whose primary object is to illuminate the role of science in regulatory proceedings, an area of decisionmaking that is often generically described as "science policy."[9] Contributions to this area of research emphasize the interplay of facts and values in public policies dealing with technological hazards. They also project a considerably less sanguine view of the power of science to influence and rationalize policy than was espoused by the contributors to Golden's volume.

The nature of policy-relevant science and its relationship to policy have been most instructively explored in a cluster of studies focusing on the U.S. government's efforts to regulate carcinogens in the 1970s and 1980s. In one of the few book-length treatments of the subject, Mark Rushefsky, a political scientist, noted that scientific uncertainty is a resource that can be mobilized by regulators and other actors in their efforts to influence policy.[10] In his account of the evolution of federal cancer policy, Rushefsky argued that competing interest groups use both knowledge and gaps in knowledge for instrumental purposes, specifically, to shape risk-assessment guidelines consistent with their social objectives. Dwelling on the interconnections between facts and values (or science and policy) in carcinogenic risk assessment, Rushefsky joined the rank of risk analysts who assert that the scientific component of the exercise can never be wholly separated from its value component.[11]

Liora Salter's study of standard-setting controversies in Canada diverged from Rushefsky's in choosing science itself as the object of investigation; put in slightly over-simplified terms, she asked how scientific activity is affected by standard-setting rather than the reverse.[12] She argued that "mandated science," the science used for purposes of making policy, has characteristics that distinguish it, on the whole, from science generated in pure research settings. Like Rushefsky, who spoke of "regulatory science," Salter acknowledged that mandated science differs from normal science partly because of ways in which society uses the two bodies of knowledge. But other distinctive features of mandated science, in her view, must be attributed to the fact that scientific and policy considerations are closely integrated at every step in its production and use.[13] Salter suggested that the procedures used by regulatory agencies should take into account the "mixed" nature of the science used in policymaking. Her own research indicated, however, that standard-setting processes, at least in Canada, continue to take their cue from an idealized picture of science that ignores its links to policy.

If science in the policy setting is always colored by values, then what role should scientists, who are professionally committed to impartiality, expect to play in decisionmaking? Joel Primack and Frank von Hippel, two early contributors to the field of science policy analysis, believed that the correct response was for more scientists to inject their own political values into science, but from a consciously environmental and public health perspective. Government agencies, they argued, had frequently misused, ignored, or concealed the opinions of their scientific advisers on important questions of science and technology policy.[14] To counteract these abuses, Primack and von Hippel urged the development of a stronger "public interest science" movement dedicated to exposing the uncertainties and assumptions buried in conventional expert assessments of scientific and technological developments. They also advocated more open advisory procedures, a recommendation that has since been widely implemented in the United States.

A later and less polemical work, Ted Greenwood's study of decisionmaking in EPA and OSHA, debunked the common complaint that agencies are scientifically incompetent and sought to explain perceived agency failures in terms of institutional factors.[15] Vulnerability, Greenwood suggested, is a bigger problem for U.S. regulators than incompetence. When key facts are unknown, regulatory agencies have to act on the basis of discretion rather than certain knowledge, thereby undermining the legitimacy of an administrative system that is, on the face of it, firmly committed to rational, nonarbitrary decisionmaking.[16] Greenwood suggested that more frequent consultation with advisory committees might increase the apparent competence of regulatory agencies while usefully reducing their scope for discretionary action.

Greenwood's conclusions about discretion are shared by most commentators on science policy in the regulatory agencies. It is now widely recognized that the questions regulators need to ask of science cannot in many instances be adequately answered by science.[17] There is also general agreement that, in the absence of sufficient hard evidence, decisions have to be made on the basis of available facts supplemented by a large measure of judgment.[18] Nevertheless, there is an unspoken presumption in many of the aforementioned works that better scientific characterization of a problem will lead to better policy. The validity of this basic assumption, however, has also begun to be questioned. For instance, although political conflict may be promoted and sustained by scientific uncertainty, it is by no means safe to assume that reducing uncertainty automatically reduces conflict. This is the intriguing conclusion that emerged from a study of safety standards for formaldehyde and benzene (both suspected carcinogens) by John Graham, Laura Green, and Marc Roberts.[19] Their account of the way scientific debates about these two substances matured over time showed that advances in scientific knowledge do not predictably correlate with reductions or increases in policy conflict.

A possible explanation for this anomaly can be found in a number of works comparing U.S. cancer policies with those of other countries. These analyses suggest that the formal and adversarial style of American regulatory decisionmaking highlights uncertainty, polarizes scientific opinion, and prevents efficient resolution of disputes about risk.[20] Far from promoting consensus, knowledge fed into such a process risks being fractured along existing lines of discord.

Similar observations drawn from other regulatory controversies led David Collingridge and Colin Reeve, two British analysts of technology policy, to form a sweepingly negative conclusion about the capacity of scientific knowledge to advance rational policymaking.[21] Science, they asserted, always encounters an under-critical or an over-critical environment when it is linked to policy; in either case, the impact of science on policy is negligible. In the under-critical model, a policy consensus exists before new research is undertaken, ensuring too easy reception of scientific claims that appear to support the policy. In the over-critical model, by contrast, political adversaries are sharply divided and scientific claims are subjected to heightened scrutiny by experts from rival camps. The result, most often observable in the U.S. regulatory process, is endless technical debate.

The theme that emerges most forcefully from these studies is that scientific uncertainty and the pressures of decisionmaking lead to a forced marriage between science and politics. Guidelines for cancer risk assessment are a typical product of this unnatural union, an unstable policy instrument in which the balance of scientific and political considerations can disintegrate at any moment as a result of changes in either knowledge or politics.

Strangely absent from the literature, however, is the puzzle this poses for scientific advisory committees attached to regulatory agencies. If the scientific claims that these bodies are asked to evaluate are uncertain, insufficient, and inherently mixed with policy, then how can advisers selected for their technical expertise and political neutrality possibly certify them as valid science? Alternatively, if regulatory advisers invariably become part of a hybrid sociotechnical process, as most of the literature suggests, then how can they maintain their authority as neutral experts, especially when challenged in the media or the courts? Finally, if Collingridge and Reeve are right, then are we not forced to conclude that scientific advice is at best simply irrelevant to policy? Published accounts of science in policy thus deepen the paradox of advice and legitimacy, for by questioning whether technical advisers can ever be dispassionate, decisive, or value-neutral, they cut at the roots of the conventional justification for scientific advice.

Expertise and Trust

Assessments of the place of experts in the American administrative process are further complicated by this society's persistent ambivalence about the degree to which technocratic values should constrain the exercise of political choice. Like Yaron Ezrahi's "pragmatic rationalist,"[22] many Americans are persuaded that even the most technical policy decisions require a judicious mixture of scientific and nonscientific judgment, and there is a concomitant fear of letting experts usurp that part of decisionmaking which should be truly political. Yet an alternative view—that components of decisionmaking requiring specialized knowledge should be depoliticized and left to experts—continues to reassert itself in American politics. The technocracy movement, which flowered for a brief period between the two World Wars, gave perhaps the most extreme embodiment to these views,[23] but support for skill-centered forms of policymaking, especially on matters concerning science and technology, remains a force to be reckoned with in modern times.[24]

The oscillation between deference and skepticism toward experts can be observed at almost every stage in the rise of the modern administrative state. By the end of the nineteenth century, the notion of administrative specialization had come into vogue and it was clear to many observers that persons wielding regulatory power needed particularized experience in order to carry out their delegated responsibilities. Woodrow Wilson, for example, approved expert agencies in general terms, recognizing that regulatory tasks require "not a little wisdom, knowledge, and experience" and that "such things must be studied in order to be well done."[25]

The idea that administrators might require technical knowledge in addition to experience was perhaps slower to gain ground, but also soon found adherents among progressive thinkers. Thus, Leonard White in 1926 wrote glowingly of the government's dependence on technical experts:

> So we discover in the administrative service one official who knows all that can be known about the control of water-borne diseases, another who has at his fingertips the substance of all available information on wheat rust, and another who cannot be "stumped" on appropriations for the national park service. These men are not merely useful to legislators overwhelmed by the increasing flood of bills; they are simply indispensable. They are the government.[26]

But not everyone was equally sanguine:

> The expert knows his stuff. Society needs him, and must have him more and more as man's technical knowledge becomes more and more extensive. But history shows us that the common man is a better judge of his own needs in the long run than any cult of experts.[27]

A more elaborate critique of experts in government developed in connection with studies of the independent regulatory commissions in the 1930s and 1940s. The Brownlow Committee report of 1937 expressed misgivings about the legitimacy of these institutions and deplored the tendency of Congress to set up powerful policymaking bodies outside the established executive departments. Characterizing these agencies as an irresponsible "headless 'fourth branch' of the Government," the Brownlow Committee recommended that their functions be redistributed to normal regulatory agencies "set up, not in a governmental vacuum outside the executive departments, but within a department."[28]

One of the strongest arguments in support of independent commissions was their alleged capacity to attract high-caliber experts to handle the tasks of regulation. In his classic study of the commissions, however, Marver Bernstein attacked this way of thinking as basically misguided. In Bernstein's view, the training and experience of experts were more likely to predispose them to myopia and bureaucratic inflexibility than to serving a broad conception of the public interest. Expertness, he suggested, would be of significant value to regulators only when all of the following conditions were met:

> (a) the scope of the problem is narrow; (b) the task of collecting data and analyzing facts is difficult and complex; (c) discretion is severely

limited; (d) the task involves the application of settled policy to regulatory situations and does not concern the formulation of basic regulatory policy; and (e) Congress has defined the public interest with sufficient clarity to guide the direction and content of public policy.[29]

Bernstein argued that these preconditions were rarely found in the work environments of most regulatory commissions. As will be seen, except for the growing difficulty of "collecting data and analyzing facts," the situation is not fundamentally different in most programs of contemporary social regulation. Ironically, as well, the increased complexity of fact-finding appears, if anything, to have diminished the authority of experts in policymaking.

The harnessing of scientific knowledge to military ends in both world wars, culminating in the nuclear bomb and the threat of global destruction, prompted darker concerns about fostering too close a fellowship among science, technology, and government. President Eisenhower's farewell warning against letting public policy "become the captive of a scientific-technological elite" was directed primarily against the military,[30] but it has resonated with later generations worried about the power of big corporations and the use of scientific expertise as a screen for activities that threaten public health, human dignity, and the natural environment.

The prospect of relinquishing any significant share of political authority to experts also goes against the grain in a society where the Jeffersonian ideal of democracy still finds ready public support. When policymakers who are not scientists ruefully refer to the cult of "doctor worship,"[31] they are merely voicing the popular conviction that decisions cannot be wholly legitimate if they are comprehensible only to the initiated. These are precisely the considerations that led to the wholesale opening up of the American administrative process to public scrutiny and participation in the early 1970s. David Bazelon, former Chief Judge of the Court of Appeals for the District of Columbia Circuit and an ardent defender of open decisionmaking, spoke for many when he argued that public supervision, not the myth of "disinterested expertise," was the key to responsible policies about technological risk.[32]

But skepticism about science and scientists seems capable of coexisting peacefully, if somewhat uneasily, with continued public confidence in policy development by experts. Derek Price, the historian who popularized the concept of "big science," was not out of tune with contemporary opinion when he wrote in 1963 that "the increased status of scientists and scientific work makes them increasingly vital to the state and places the state increasingly in the position of putting technical decisions in technical hands."[33]

As if bearing out Price's optimism about experts, Jon Miller's work on public attitudes toward science policy provides suggestive evidence that the majority of citizens would rather entrust technical decisions to specialized committees than to any other form of institutional authority. Miller reports on a 1979 survey that asked respondents to identify the groups or individuals they considered most qualified to resolve issues in three policy areas: space exploration, nuclear power, food additive regulation. In all three, "scientists and engineers who specialize in this area" were most frequently designated as the preferred decisionmakers by both the "attentive" and the "nonattentive" public. In contrast, the second most preferred decisionmaker varied by issue area (federal regulatory agency for food additives; citizen referendum for nuclear power).[34]

The Contingency of Knowledge

The idealized picture of science from which the advisory process has traditionally drawn its authority has come under attack not only from political scientists and policy analysts but from a thriving area of scholarship that has abandoned the notion of science as a representation of objective reality in favor of a closer inquiry into the social processes by which scientific knowledge is produced or "constructed." Central to this body of work is the attempt to understand what makes scientists accept some claims as better than others, given that confirmation is not to be had through simple appeal to the external world. Ethnographic studies of laboratories, historical accounts of the rise and demise of particular scientific theories, and investigations of public controversies involving science and technology have all provided fruitful insights into the processes by which an image of reality gains acceptance as the real thing. This body of work suggests that science, far from being part of the solution, may in fact be part of the problem that confronts the makers of science policy.

While this book is not the place for an extended discursus on the sociology of science, three major findings from this field must be taken into account in any serious discussion of scientific advising. The first is the observation that scientific "facts" are, for the most part, socially constructed.[35] We regard a particular factual claim as true not because it accurately reflects what is out there in nature, but because it has been certified as true by those who are considered competent to pass upon the truth and falsity of that kind of claim. Social construction begins in the laboratory, where most scientific claims originate,[36] but may reach out to include wider communities, including the news media and the lay public. Particularly relevant to the task of advisory committees are studies that show how claims related to technological risk are socially constructed and how players with different stakes in technical controversies arrive at different constructions of scientific reality.[37]

If scientific claims are constructed, then it follows that they can equally be *deconstructed,* thereby losing their factual status through reidentification of their social origins. In the formulation used by Bruno Latour and Steve Woolgar, it should be possible under appropriate circumstances to melt "reality" back into its constituent statements, "the conditions of production of which are once again made explicit."[38] Studies of the U.S. regulatory process suggest that it provides a particularly fertile environment for deconstruction, since major stakeholders have an interest in tearing down one another's version of constructed reality.[39] A primary concern of this book is to investigate how scientific advice affects both the construction and the deconstruction of claims in the regulatory process.

A second and related way in which sociology of science impinges on the subject matter of this book is by challenging the notion that scientific facts are tested and established with reference to objective criteria of validity. Thomas Kuhn's classic account of scientific change took an important step in this direction by positing that accepted scientific activity in any period is merely that which conforms to the prevailing paradigm.[40] It is the paradigm, rather than any feature of the natural world, that defines what problems are worth solving and shapes scientists' expectations of what they are likely to see when they investigate nature.

Among the sociological studies that have elaborated upon the implications of Kuhn's original analysis, Harry Collins's work on replication is particularly relevant to problems that face advisory committees. Collins identified a phenomenon that he called "experimenters' regress."[41] This is the circularity that sets in when there is no universally accepted "objective" criterion for determining whether an experiment has been competently performed. In such situations, debates about the quality of the experiment cannot be separated from debates about its output; belief in the former requires belief in the latter, although the reality of the result ostensibly cannot be established without the properly conducted experiment. When an objective or "scientific" test of experimental quality is unavailable, Collins's data suggest that scientists freely turn to nonscientific criteria of excellence, such as faith in the experimenter's honesty, the size and prestige of the laboratory, and even personal qualities like nationality or professional group affiliations.[42] A theme that will be developed throughout this book is that the embedding of science in political frameworks exacerbates these tendencies toward personalizing issues of experimental quality.

If the most important project of sociology of science in recent years has been to expose the contingent and relativistic character of knowledge, a second and scarcely less important project has been to illuminate how science nonetheless succeeds in acquiring and maintaining cognitive authority in a distrustful world. Research on the latter project provides the third significant point of contact between sociology of science and the concerns of this book, for it suggests how scientific advisory committees are able to preserve the appearance of authority even in the face of uncertainty and political conflict.

One of the most frequent strategies used by scientists to enhance their authority is what the sociologist Thomas Gieryn has referred to as "boundary work."[43] Whether they are engaged in building professional communities, defining and excluding nonmembers, competing for resources, or asserting their autonomy against external controls, scientists use a variety of boundary-defining strategies to establish who is in and who is out of relevant peer groups and networks of prestige or authority. The most consequential—and exclusionary—of all possible boundaries is that between "science" and other systems of cognitive authority, such as religion or law. When an area of intellectual activity is tagged with the label "science," people who are not scientists are *de facto* barred from having any say about its substance; correspondingly, to label something "not science" is to denude it of cognitive authority. As we shall see throughout this book, this feature of boundary work assumes tremendous importance in debates over regulatory science, which almost by definition straddles the dividing line between science and policy. Participants in the regulatory process often try to gain control of key issues by changing their characterization from science to policy or from policy to science.

The Reform Debate

The complexity and subtlety revealed by the accumulating literature on science policy and sociology of science have exerted surprisingly little impact on proposals for improving the quality of science-based regulation. Two rather simplistic conceptions of the place of science in public decisionmaking continue to dominate the policy literature, entailing diametrically opposite prescriptions concerning the role of advisory committees.

According to one viewpoint, the technical incompetence of the bureaucracy is the most significant barrier to making the "right" decisions at the frontiers of scientific knowledge. Regulators are seen as insufficiently expert at distinguishing "good" from "bad" science and as insensitive to the standards by which the scientific community evaluates evidence. Adherents of the bureaucratic incompetence school recognize, but deplore, the amount of discretion agencies enjoy in evaluating science. Such broad delegations, in their view, simply offer decisionmakers a carte blanche to disregard information incompatible with their political goals and to formulate science policy decisions that cannot pass muster with qualified scientists.

A very different view of the problem prevails among traditionally proregulation interests—the environmental, labor, and consumer movements. Their allegations that administrative agencies misuse science usually rest on a perception that regulators are not sensitive enough to the legislative policies and social values that should guide the evaluation of complex and uncertain data. Seen from this perspective, bias in scientific assessments is most often the result either of conscious deception by industrial experts or of an uncritical acceptance of industry's viewpoint by agency officials. In either case, the problem can be seen in terms of the classic "capture" paradigm: an agency grown too close to those it seeks to regulate tends to accept unquestioningly the self-serving view of risk advanced by the regulated interests and their hired experts.

These disparate analyses of the causes of failure in science policy are associated with equally divergent philosophies about the desirable directions for reform. The "technocratic" view consistently favored by commercial and industrial interests holds that the solution is to get more and better science into decisions.[44] The recommended way of achieving this end is to expand the role of the expert community in decisionmaking. Proposals for accomplishing this objective include the separation of scientific and political decision-making, in part by conferring more authority on scientific advisory bodies. One refrain heard with increasing frequency in recent years is that agencies should ensure that their decisions are peer-reviewed in accordance with the normal practices of the scientific community. Other proposals include removing certain types of decisions (for example, the development of risk-assessment guidelines) from the control of the agencies altogether and delegating them to such scientifically irreproachable institutions as committees of the National Academy of Sciences.

The "democratic" critique of science policy, by contrast, holds that the primary problem is the failure of the regulatory agencies to incorporate a full enough range of values into their decisionmaking. Representatives of this point of view generally ask for mechanisms that will broaden the participatory base of agency action. If technical advisory committees must be used, advocates of populist reform urge that the membership of such bodies be diversified to include more than narrowly technical viewpoints. Such critics also emphasize the need for nonscientific modes of accountability: open decisionmaking procedures, advance publication of decisionmaking guidelines, and judicial review.

In short, the positions adopted by the major interest groups with a stake in regulation approximate the two idealistic formulas that Don Price outlined for bridging "the spectrum from truth to power."[45] Today's technocratic critics of regulation eagerly accept the notion that the scientific community should organize itself so as to play a more active part in the formulation of social policy. Indeed, the technocratic critique of science policy consistently identifies insufficient consultation with the scientific community as a principal cause of regulatory failure. The democratic alternative of broader interest representation, in contrast, rests on the presumption that the average citizen can be sufficiently educated on technical issues to play an informed role in the policy process.

I have suggested in this chapter that neither the technocratic nor the democratic model accurately captures what is at stake in decisions that are at once scientific and political. The notion that the scientific component of decisionmaking can be separated from the political and entrusted to independent experts has effectively been dismantled by recent contributions to the political and social studies of science. With the accumulation of evidence that "truth" in science is inseparable from power, the idea that scientists can speak truth to power in a value-free manner has emerged as a myth without correlates in reality. At the same time, as the following chapters will demonstrate, it has become clear that broad citizen participation alone cannot legitimate decisions that do not command the respect of the scientific community.

In order to prove genuinely useful, proposals to improve the use of science in the regulatory process have to be informed by an accurate knowledge of the internal dynamics of both science and regulation. The regulatory ideology of the early 1970s was marked by a sometimes naive faith in the power of American institutions to identify and control technological insults to public health and the environment. Subsequent years have seen a retreat by Congress, the courts, and the agencies themselves from an excess of optimism. There is a recognition that the economy is not infinitely robust, that knowledge is imperfect, and that some risks may have to be tolerated in order to encourage innovation and secure the progressive benefits of technology. Numerous new approaches to regulation—the "bubble" policy, the use of offsets, *de minimis* risk analysis, right-to-know policies—testify to a more cautious, pragmatic, and incremental definition of objectives than was prevalent a decade ago. A similar caution has to mark any attempt to improve the framework of scientific advice-giving. However rhetorically appealing it may be, no simple formula for injecting expert opinion into public policy holds much promise of success.

An Alternative Approach

In examining the interaction between expert committees and agencies, I attempt to break away from the largely ahistorical and case-oriented literature on science policy. Although the chapters on scientific advisory mechanisms contain material on current policy controversies, I make an effort to root them in a deeper historical setting. Individual regulatory proceedings are presented as stories with a temporal dimension corresponding to changes in national politics and scientific knowledge. Finally, as this opening chapter indicates, my chosen approach is interdisciplinary. It seeks to incorporate insights not only from the "expected" fields of law, political science, and policy analysis, but also from areas of scholarship that are more particularly concerned with understanding the nature of scientific knowledge and its relation to political power.

In line with these objectives, the first four chapters of the book criticize and set aside the two prevailing models by which science is legitimated in regulatory policymaking. In particular, Chapter 2 gives an account of four controversies that helped sharpen public awareness of problems in the field of science policy and revealed serious institutional deficiencies in the production and use of regulatory science. The chapter reviews the reform proposals that grew out of these four incidents and argues that the controversies in fact carried a more ambiguous message than is evident from these proposals. Chapters 3 and 4 describe, respectively, the shortcomings of the democratic and technocratic models of incorporating science into policy. Chapter 3 analyzes the primary nontechnocratic methods of securing accountability in science policy decisions: judicial review and open decisionmaking. The limitations of these techniques underscore the continuing need for scientific advisory mechanisms in the regulatory process. Chapter 4 contrasts the science used in policymaking ("regulatory" or "mandated" science) with "research" science and uses the analytical literature on science fraud to argue that many of the problems of accountability identified in the former are actually encountered, albeit in more attenuated form, within the latter.

Chapters 5 to 9 present empirical data about the way advisory bodies are used by two of the most intensive consumers of science among the federal agencies: EPA and FDA. Specifically, Chapter 5 describes the role of EPA's agency-wide Science Advisory Board, while Chapters 6 and 7, respectively, look at two committees attached to self-contained regulatory programs under the Clean Air Act and the Federal Insecticide, Fungicide, and Rodenticide Act. Chapter 8 assesses a variety of advisory committee structures used by FDA in its decisionmaking on pharmaceutical drugs and food additives. The objective of Chapter 9 is to illustrate how expert advisory systems function when scientific decisionmaking has to accommodate changes in both politics and knowledge. The two cases analyzed in this chapter are the development of guidelines for cancer risk assessment and the regulation of formaldehyde.

In Chapter 10, the focus shifts to mechanisms other than advisory committees that have succeeded, to varying degrees, in shifting regulatory decisionmaking toward the technocratic model. The chapter also evaluates the prospects for generalizing these approaches to other areas of policymaking. Finally, Chapter 11 presents a revised and enriched picture of the way science interacts with politics in the regulatory process and indicates how the provision of scientific and technical advice to regulatory agencies can be improved and made more effective.

2

Flawed Decisions

Between 1975 and 1980, a number of extremely visible controversies over chemicals in food, the environment, and the workplace rekindled public debate about the need for a closer working relationship between regulators and the independent scientific community. In most cases, events followed a distressingly similar pattern. A regulatory agency would propose to ban or restrict a chemical on the basis of an in-house scientific study or risk assessment. The agency's findings would be challenged by outside experts, and further review of the data would lead to a modification of the agency's original scientific position. The embarrassed agency would have to withdraw its policy initiative, with predictable damage to its scientific and political reputation. Collectively, these incidents fostered the impression that agencies could not be trusted to use science in a responsible manner unless they were supervised by a more neutral scientific authority, such as an advisory committee.

Although these controversies seemed on their face to be about science, the alignment of parties on either side generally conformed to basic political and ideological cleavages between pro- and antiregulation interests in American society. Industry groups and conservative policy analysts typically led the charge against the regulatory agencies, accusing them of subverting science in order to placate antagonists of technology and big business. Edith Efron's skillful, though polemical, exposition of these views in *The Apocalyptics*[1] rallied the chemical industry in much the same way that Rachel Carson and Samuel Epstein had galvanized environmentalists a few years earlier.[2] Efron argued that federal cancer policies in the 1970s displayed a systematically proregulatory bias and were the work of activist scientists bent on a wholesale transformation of the American economic system. Other critics stopped short of attributing willful misbehavior to government experts, but argued that agencies were getting away with shoddy science because of deficiencies in the existing mechanisms of quality control. These analyses spurred a sharp rise in the demand for scientific peer review of regulatory proposals.

There emerged, however, a countervailing position with implications that were much less hostile to the agencies. According to this view—generally subscribed to by environmentalists and labor and consumer groups—it was entirely legitimate for agency decisions about the quality or sufficiency of risk information to take account of social policy considerations. Judgments concerning what to do about uncertainty, in particular, had to reflect the will of Congress, even if this led to differences of opinion between the regulatory agencies and experts speaking for other players in the regulatory process. For instance, a strongly precautionary statute, such as the Delaney clause[3] or the Clean Air Act, might require agencies to act on the basis of scientific information that did not measure up to standards of causation accepted by research scientists and industry experts. Similarly, an agency might espouse a conservative risk assessment as prudent public policy even if some nonagency scientists characterized this approach as "bad science."[4]

The four controversies described in this chapter conformed to the canonical pattern of science policy disputes outlined here, though they were perhaps atypical in the amount of publicity they generated. In each case, powerful voices representing both science and industry identified inadequate "peer review" as an important cause of the agency's scientific problems and argued—in part successfully—that legislative amendments were needed to force more consultation between agencies and outside experts. At the same time, these cases left unresolved some of the basic questions concerning the interpretation of science in the policy environment. In cases of disagreement over regulatory science, who should be designated as final arbiters of quality: the agency, the courts, an independent scientific body? Can there ever be an objective basis for determining what is "bad science" (that is, weak, inconsistent, methodologically flawed science)? In any event, does "bad science" inevitably make for bad policy; and conversely, can good policy ever be made in the absence of "good science"?[5]

Nitrites

The Food and Drug Administration in 1978 was still recuperating from the political debacle over saccharin, when it confronted a major new scientific controversy over nitrites added to food. Its proposal to ban the artificial sweetener as a suspected carcinogen had encountered immense public resistance. Few could understand why so important and useful an additive should be eliminated on the basis of effects in rats fed the equivalent of 800 cans of diet soda a day over a lifetime.[6] Was FDA wholly insensitive to the needs of dieters and diabetics? Congress, at least, had proved more responsive to the popular will; it enacted a special law to keep saccharin on the market, in the process dealing a severe blow to FDA's prestige.[7]

To an agency trying to repair its tarnished image, the nitrite case came as a most unwelcome surprise. Paul Newberne, a scientist at the Massachusetts Institute of Technology (MIT), reported on the strength of a three-year FDA-sponsored rodent feeding study that nitrites caused an increase in lymphatic cancer and in the number of "precancerous lesions" observed in the test animals.[8] His study design appeared to rule out any possibility that these effects were actually caused by nitrosamines, a class of known human carcinogens formed when nitrites combine with amines. If Newberne's findings were correct, then, pursuant to the Delaney clause, some of the major uses of nitrites would have to be banned. The economic consequences promised to be devastating, since nitrites were used as a preservative in some 7 percent of the U.S. food supply (mainly poultry, fish, and meat).[9] The public health implications of a nitrite ban were also troubling. Nitrite treatment of meat and fish had reduced the risk of botulism in the United States to near zero, and no effective substitutes were immediately to be found.

Aware of the high stakes and the need for caution, Donald Kennedy, then Commissioner of FDA, opened discussions with the Department of Agriculture (USDA) to develop a joint policy on nitrites. Within a few months of receiving Newberne's report, the two agencies agreed in principle that they should call for a gradual phasing out of all uses of nitrites.[10] FDA took great pains not to announce its decision prematurely or to press for instant action, but the agency's strategy ran into unexpected obstacles when Joseph Califano, then Secretary of the Department of Health, Education and Welfare (the predecessor to HHS), bypassed FDA and consulted the office of the Attorney General about the legality of phasing out nitrites. Disagreeing with FDA's lawyers, the Attorney General indicated that a gradual phase-out would not, in his opinion, be in compliance with the Delaney clause and other related provisions of the Food, Drug, and Cosmetic Act.[11] If nitrites did indeed induce cancer in animals, then the agency's only choice was to enact an immediate ban. A frustrated Califano turned to President Carter for help in getting Congress to approve FDA's phase-out plan for the suspect compound.[12]

At this point, the nitrite story took an even more surprising turn. FDA announced in 1980 that it was withdrawing its earlier statements of concern about the additive's possible carcinogenicity.[13] A thoroughgoing review of Newberne's study had persuaded the agency that nitrites alone did not cause cancer in exposed animals. As the full story of this turnaround became public, FDA found itself embroiled in renewed charges of reacting in hasty and ill-considered fashion to scientific information that did not stand up to responsible professional scrutiny.

Scientists asked to comment on the MIT study had, it now appeared, expressed doubts from the start about the adequacy of Newberne's data, the quality of his statistical analysis, and the correctness of his interpretation of the pathological data. Questions had also been raised about animal handling practices in Newberne's laboratory when FDA inspectors visited it during the course of the experiment. In order to set these questions to rest, FDA had the study intensively reviewed by its own working group on nitrites, scientists from three other governmental agencies, and a committee of experts from Universities Associated for Research and Education in Pathology (UAREP), an independent consulting group. To the agency's consternation, the conclusions of these reviewing bodies appeared to exonerate nitrites completely. In particular, the UAREP scientists determined that Newberne had mistakenly diagnosed as lymphomas a rare form of cancer that could not have been produced by nitrites, and that the so-called precancerous lesions observed by Newberne were not real tumor precursors.

How could a study with so many problems create such alarm and tie up the agency's attention and resources for two years? An explanation that found immediate favor was inadequate peer review. According to one reading of the story, Commissioner Kennedy, a distinguished biologist, went wrong in acting on his own intuition about nitrites, without subjecting his views to the discipline of FDA's normal review process for scientific studies. Kennedy had consulted a special task force of a dozen FDA lawyers and scientists about the Newberne study, but had not fully involved the toxicology staff in the Bureau of Foods in his final decision.[14] Nonetheless, he assured Califano, "We know more than enough about the Newberne study to be convinced that it is well done and strongly supports the hypothesis that nitrites are carcinogenic *per se*."[15] Was this perhaps a case of expert self-deception, with a trained scientist "seeing" a certain result simply because he expected to see it?[16] Would Kennedy's conviction have been shaken if he had consulted the FDA experts most familiar with nitrites? The president of the American Meat Institute certainly saw the events in this light. Denouncing FDA's entire initiative on nitrites as "ill advised," he asserted that "peer review of studies of this type should be an essential element of the process."[17]

But there were some ambiguities in the record that seemed to speak in favor of a more shaded verdict. To start with, Newberne himself was not completely persuaded by the UAREP review. With regard to the pathological data, for example, he suggested that the disagreement between the UAREP experts and himself was in part terminological; the tissues, he believed, showed genuine nitrite-linked changes, no matter how one chose to label them.[18] Kennedy also was convinced that Newberne's study confirmed some harmful effects on the lymphatic system induced by nitrites, even if these effects did not amount to malignant tissue change.[19] If one accepted these interpretations of the data, Kennedy's decision to phase out nitrites over time began to look more like a legitimate attempt to err on the side of caution while reading borderline science for the protection of public health. Prudence was perhaps particularly warranted in this case because nitrites are also believed to combine with free amines to form the indubitably carcinogenic nitrosamines. Indeed, Jere Goyan, Kennedy's successor at FDA, noted that the agency's retreat from Newberne's findings did not give nitrites a clean bill of health.[20] However, these subtleties were on the whole overshadowed by FDA's public dismissal of Newberne's study, and the nitrite episode entered the annals of regulatory case histories as supporting stronger peer review of policy-relevant science.

2,4,5-T

The herbicide 2,4,5-T was recognized as a possible environmental hazard years before it became a focus of controversy at the Environmental Protection Agency. Commercial preparations of the compound were known to be contaminated with dioxin, a potent toxic chemical associated with a wide variety of adverse health effects, including teratogenicity, carcinogenicity, and a skin disease called chloracne.

Two catastrophic events propelled 2,4,5-T to the forefront of public awareness in the late 1970s. During the Vietnam war, American troops used massive quantities of the defoliant Agent Orange, a herbicide containing 2,4,5-T, in their effort to strip forests and expose the hiding places of enemy guerrillas. After the war, there were reports of unusually high rates of birth defects, cancers, and neurophysiological disorders among American soldiers and Vietnamese civilians exposed to Agent Orange. Fears about 2,4,5-T were exacerbated by a pesticide-plant explosion in Seveso, Italy, which released a cloud of dioxin and necessitated evacuation of much of the surrounding countryside.[21]

By this time, most uses of 2,4,5-T on food crops were already banned in the United States on the strength of laboratory tests linking the compound to birth defects in animals. EPA, however, became concerned that the compound's remaining uses might pose unreasonable health risks, and epidemiological studies were undertaken in several countries, including the United States, to determine the risks of continued environmental exposure. One study that appeared to confirm the agency's fears was a survey carried out for EPA by researchers from various universities in Oregon's Alsea basin, the site of regular annual 2,4,5-T spraying.[22] The study found an apparent correlation between the spraying activities and peaks in the regional miscarriage rates during a six-year period. Responding to the study, EPA promptly suspended further use of 2,4,5-T on forests and rights of way.

Both the study and EPA's reaction to it quickly came under fire from industry. Etcyl Blair of Dow Chemical, a major producer of 2,4,5-T, decried EPA's action as "an example of government at its worst—basing a hasty product suspension on data which have not been subjected to scientific review."[23] Although EPA's decision survived a challenge in court, the judge, too, was troubled by the apparent weaknesses in the Alsea study.[24] He ultimately upheld the suspension, but with great reluctance and only because he did not wish to substitute his own judgment for the agency's. Subsequent scientific review of the Alsea basin study seemed to confirm the misgivings of the skeptics. One member of the Scientific Advisory Panel (SAP), a group of independent scientists advising EPA's Office of Pesticide Programs, stated at an open panel meeting that flaws in the Oregon study raised doubts about its major conclusions.[25] Among the problems noted by scientists both in the United States and abroad was the absence of data on the levels at which pregnant women were exposed or on whether exposure occurred at a stage in fetal development consistent with the reported miscarriages.[26] Convinced that it could not rely on the Alsea study to ban 2,4,5-T, EPA began negotiating with Dow to reach a voluntary agreement with respect to the compound's remaining uses.

Despite widespread scientific repudiation of the epidemiological study carried out at Alsea, events conspired to keep 2,4,5-T and dioxin in the news. Vietnam veterans brought a large class-action suit against the manufacturers of Agent Orange in a federal district court in New York.[27] A spectacular case of dioxin contamination at Times Beach, Missouri, forced the federal government to buy up homes in the polluted area.[28] Finally, in 1983 Dow unexpectedly requested EPA to cancel the remaining registrations for 2,4,5-T, claiming that it was no longer financially worthwhile to continue defending the compound's safety.[29] EPA's precautionary philosophy—based largely on information about high-dose exposures to dioxin—thus prevailed in the end, even though the agency was never able to muster a scientifically convincing case that low-dose environmental exposures posed a significant risk to human health.

The other lasting legacy of the 2,4,5-T controversy was to tighten up peer-review procedures in EPA's pesticides program. Congress had already determined in 1975 that the agency should be legally required to consult with external scientific advisers in evaluating the environmental and health risks posed by pesticides. The 1975 amendments to the Federal Insecticide, Fungicide, and Rodenticide Act (FIFRA) enlarged SAP's duties to include the review of suspension decisions. While the statute still permits EPA to order "immediate suspension" of a pesticide without SAP's approval, the decision must "promptly" be submitted to the advisory committee for scientific validation. The impact of such review on subsequent decisionmaking about pesticides is the subject of Chapter 7.

Love Canal

The pollution and public health disaster at Love Canal gave rise to one of the most intense and convoluted scientific controversies in recent memory, with ramifications that did much to reinforce the appearance of agency incompetence. Science entered the picture in this case after regulatory activities were already under way. EPA had identified Hooker Chemicals, a New York-based company, as the party responsible for cleaning up several toxic chemical dumpsites in the Niagara

Falls area. One of the worst-affected locations was Love Canal, a residential neighborhood built on a former chemical dump used by Hooker. For some time, residents of the area had complained of unusually high incidences of a variety of illnesses, including cancers and birth defects, which in their view were linked to toxic chemicals seeping into their homes from waste drums buried in the Canal. In an effort to provide formal scientific corroboration for these anecdotal reports—and to strengthen its legal position against Hooker—EPA commissioned from a private consulting firm, Biogenics Corporation, a study of chromosomal aberrations among residents of the contaminated neighborhood.[30] Dante Picciano, the head of Biogenics, designed and conducted the study in collaboration with Beverly Paigen of Roswell Park Memorial Institute, who also collected the blood samples from the participants in the study. Picciano was led to believe that EPA wanted results as soon as possible,[31] and the study was geared to meet this need.

Working under both time and resource constraints, Picciano reported to EPA in early 1980 that he had found an unusually large number of chromosomal abnormalities among 36 study participants, including breakages and a type of malformation which he described as "supernumerary acentric chromosomes."[32] Before EPA could analyze these results any further, however, they were leaked to the press, and proliferating news reports of chromosomal damage at Love Canal spread panic among already anxious area residents. Yet as soon as Picciano's results were made public, they were bitterly contested by experts within and outside government. The study was reviewed by two separate and distinguished panels: a group of eight scientists led by David Rall, the director of the National Institute for Environmental Health Sciences (NIEHS), and another committee headed by Roy Albert, a longtime scientific adviser to EPA.[33] Both groups concluded that the Picciano study could not be relied upon in drawing either positive or negative conclusions about the cytogenetic condition of Love Canal residents. Criticism focused on four aspects of the study: the poor quality of the cytogenetic preparations, the questionable interpretation of the abnormalities, the absence of simultaneous controls, and the possible irrelevance of the apparent chromosomal aberrations to actual health effects. Particularly damaging was the Albert panel's observation that the so-called supernumerary acentric chromosomes seemed to exist only in Picciano's imagination.[34] A third review carried out by a blue-ribbon panel appointed by Governor Hugh Carey of New York seemed to clinch the negative verdict. Inadequacies in the study, the panel concluded, could have been discovered through timely peer review, and it emphasized the need for agencies to seek competent scientific review of protocols and findings before making study results public.[35]

The episode of the "botched" Love Canal study thus provided strong ammunition to those who argued that science was too important to leave in the hands of regulatory agencies. At EPA, Stephen Gage, assistant administrator of the Office of Research and Development, spoke out defensively on behalf of his employer. Publicly regretting that the study had been released to the press before peer review, Gage blamed the agency lawyers and their rush to litigation for the unfortunate breakdown in EPA's established practice of peer-reviewing research protocols and research results.[36] But as in the two preceding cases, some of the details of the Love Canal incident suggested that there might be difficulties with this rather stark assessment of Picciano's and EPA's problems.

One especially suggestive aspect of the case was that experts looking at Picciano's data did not all agree on what they were seeing or why. For example, Margery Shaw, a well-known geneticist at the University of Texas, wrote a letter to *Science* saying that, unlike the Albert panel, she found no serious problems with the quality of Picciano's cytogenetic preparations and, moreover, that the photocopied slides did show a number of long, acentric fragments.[37] Picciano himself implied that the Rall review was tainted by a quite mundane and personal bias against him, noting in another letter to *Science* that the review team did not include a single scientist nominated by his own research group.[38] But the reviewer Picciano had proposed, Jack Killian, was problematic in his own right. As former colleagues at Dow Chemical, Picciano and Killian had collaborated on a study of chromosome damage among Dow workers which the company had rejected as scientifically flawed and refused to publish.[39] In a final twist, when the Rall panel failed to accept Killian as a member, Picciano refused to release his Love Canal slides, so that the panel had to render a judgment without ever seeing the raw data on which Picciano's conclusions were based.

One point on which the reviewers all agreed was that the absence of simultaneous controls rendered Picciano's study entirely inconclusive. But even this consensus was less universal and less "scientific" than it appeared on the surface. To begin with, the reviewers' reactions may have been influenced by allegations of scientific incompetence, and even chicanery, surrounding Picciano's work. Perry Gehring, Dow Chemical's director of health and environmental sciences, suggested that the historical controls used in the Love Canal study—apparently the same as those previously used in the discredited Dow study—had been doctored by Picciano and Killian to show exceptionally low rates of chromosomal damage.[40] On the other side, Picciano defended his study on pragmatic as well as methodological grounds. His failure to use simultaneous controls, he argued, was entirely a consequence of EPA's hurried approach to the study, and he blamed the agency for not following through on its initial agreement to provide 25 controls as well as 25 subjects.[41] In any event, he noted, the conclusions were cautiously formulated so as to take note of this problem. The study had explicitly acknowledged that without an appropriate control group "no unequivocal statement can be made concerning the cytogenetic results from the Love Canal residents."[42] The Albert panel further muddied the issue of controls by claiming that Picciano's results were within the "normal limits" for chromosomal damage in the general population. As Shaw pointed out, such a determination could at best have been made with reference to historical controls, since the panel had no access to concurrent control data from Love Canal.[43] How then could the Albert panel take Picciano to task for *his* reliance on historical controls?

Finally, both the Thomas panel and Gage at EPA were persuaded that the quality of EPA's decisionmaking with respect to the Picciano study was compromised by premature disclosure of the results to the media. The Thomas panel specifically recommended that in the future such disclosures should be avoided until after the completion of peer review. It is certainly arguable that such a procedure would have prevented, or at least limited, the public controversy over Picciano's findings, thus sparing EPA considerable embarrassment. But the differences among Picciano, the review panels, and Margery Shaw prompt one to ask how peer review would really have affected the quality of the science at issue. Would earlier review have made all of the scientists agree on how they should look at the data, or would it simply have weeded out "aberrant" ways of looking, in this case Picciano's (and perhaps Shaw's)?

Estimates of Occupational Cancer

The last in the quartet of case studies considered here is in some ways the most anomalous. The controversy in this case centered neither on a specific regulatory decision nor on the scientific merits of a particular study or the competence of a particular researcher or research team. Yet, although the disputed scientific exercises were not produced in furtherance of a specific policy initiative, they arguably had a more far-reaching impact on federal regulatory policy than any of the studies described above. Unusual, too, was the fact that the government's handling of science ultimately came under attack through mechanisms other than litigation or formal peer review.

The story began in the mid-1970s, when a number of scientists working for the federal government began speculating on the impact of industrialization on cancer rates in the United States. A question of particular interest was the extent to which cancer might be causally related to contact with toxic chemicals in the workplace. By 1978 an unpublished paper providing some alarming answers began circulating among federal agencies responsible for regulating chemicals. Bearing the names of nine well-known government scientists, including the directors of the National Cancer Institute (NCI) and NIEHS, the paper estimated that 20–40 percent of all cancers in the United States might be caused by occupational factors.[44] This projection, and numerous subsidiary conclusions about specific substances, created an understandable stir among federal policymakers and prompted OSHA, in particular, to start work on a generic policy for classifying and regulating occupational carcinogens.[45] In spite of its impact, however, the paper was never published, and its distinguished authors claimed neither credit nor responsibility for its conclusions.

Investigators at NCI, in the meantime, carried out related research on the overall statistics of cancer in the postwar period to see what could be said about the incidence of the disease. A paper by Earl Pollack and John Horm published in the NCI journal in 1980 concluded that the rate of cancer in the United States was rising even after making allowances for increases associated with cigarette smoking.[46] Another NCI scientist, Marvin Schneiderman, announced at a professional meeting in late 1979 that industrial exposure not only was contributing to present cancer rates but could be expected to contribute even more significantly in the future. Neither Schneiderman's paper nor the supporting data ever appeared in a refereed journal.[47] However, statements such as his fostered a growing public fear that the United States was in the grip of a "cancer epidemic."

At the same time, epidemiologists outside the circle of U.S. government scientists began raising substantial questions about the integrity of these frightening projections. Richard Peto, a highly regarded British epidemiologist, pointed out in a 1980 article that the occupational cancer estimates in the NCI-NIEHS paper were unreasonably high by any reckoning.[48] For example, the methods used in the paper would have predicted an *extra* 200,000 respiratory tract cancers from the cumulative effects of just six industrial carcinogens. The actual number of respiratory tract cancers in the United States at the time, as Peto noted, stood at just 90,000. Since most of these were attributable to cigarette smoking, the U.S. government's estimate that more than twice as many additional cancers would be induced by chemicals appeared thoroughly unrealistic. Edith Efron dealt another blow to the NCI studies by denouncing them as attempts by environmentalists to advance their agenda of spiritual and moral revival at the expense of scientific honesty.[49] In Britain, Peto proposed a similar explanation. The "apocalyptics" were successful in selling their version of science, Peto conjectured, because they supplied a picture of industrial chemicals consistent with "what many people want to believe of modern society."[50]

Such decentralized and relatively unstructured peer criticism eventually had the same effect as formal peer review in the three preceding cases. The "cancer epidemic" theory gradually lost force as a driving wheel of policy. A symptom of change was the decision of the Office of Technology Assessment (OTA) to commission a new study of the environmental causes of cancer from Richard Peto, the outspoken critic of U.S. efforts in this area, and his eminent Oxford colleague Sir Richard Doll. The two British epidemiologists concluded not only that U.S. cancer rates were not rising overall, but that occupational causes accounted for only a small percentage of the total incidence of the disease. Published in the NCI journal in 1981, their work was widely hailed as having set the record straight on cancer facts in the United States.[51]

But although many epidemiologists tacitly went along with Doll and Peto in repudiating the extreme claims of the NCI scientists, there was never any clear public adjudication of the controversy. Were U.S. experts like Schneiderman, Pollack, and Horm simply misled by political ideology, or were they operating with different assumptions and framing different questions from Doll and Peto? Was it possible, for instance, that although there was no evidence of a general "cancer epidemic," rates for certain cancer types and sites were indeed rising, as some U.S. epidemiologists believed? In the absence of a definitive resolution of these issues, questions about the links between chemicals and cancer remained unsettled and, under appropriate circumstances, ready to become again politically contentious.

The Technocratic Response

To the technocratic critics of regulation, incidents such as these carried an obvious and simple message. Faulty peer review, they concluded, was responsible for shortcomings in the way agencies were using science, and stiffer external supervision was the only way to rectify this deficiency. Numerous proposals were made for achieving this goal, and Congress, as we see below, began taking them to heart.

Separation versus Peer Review

The movement to bring more scientific accountability to regulatory decisionmaking produced two somewhat divergent policy proposals. The more radical alternative was to bring about greater institutional separation between the processes of scientific and policy analysis. Impetus for this approach derived in part from a 1978 report of the Office of Science and Technology Policy (OSTP), which recommended that different institutions should be responsible for characterizing the risks presented by chemicals (a mostly "scientific" exercise in OSTP's view) and for the analysis and design of control options.[52] The less radical approach was to leave both functions within the agencies, but to build stronger bridges between the agencies and the scientific community; regulatory peer review was the primary mechanism discussed in this connection.

To nobody's surprise, the institutional separation model received strong support from the chemical industry, which had consistently argued that regulatory agencies were abusing their power to interpret science. The American Industrial Health Council (AIHC), a chemical-industry lobbying organization, recommended in 1979 that a central expert panel be created under the umbrella of the National Academy of Sciences to assess the risks of any chemical considered for regulation.[53] The panel's findings would be advisory, but would be incorporated into the agency's decisionmaking record. An even stricter proposal was introduced into Congress by Representative William Wampler, who contemplated not only that certain risk-assessment functions would be referred to an independent body but also that the expert panel's assessments would bind the affected agencies.[54]

The arguments for separating risk assessment from the remainder of the regulatory process were explored in detail, and largely rejected, in a National Research Council (NRC) study of risk-assessment practices in the federal government.[55] Some of the reasons were purely pragmatic. The NRC report noted, for example, that it might be difficult to coordinate risk assessment with risk management, so that delays would inevitably be magnified.[56] More important, the NRC report concluded, as others have done, that risk assessment is a hybrid process combining scientific and policy judgments. Therefore, merely entrusting this function to a separate institution would not necessarily ensure that policy elements would be clearly distinguished from scientific ones.[57] Finally, the NRC expert committee did not believe that it had enough information about the risk-assessment strategies used by different agencies to feel confident about recommending a single institutional format for all the agencies.[58]

A more productive approach, in the NRC committee's view, was to require the agencies to strengthen their procedures for peer-reviewing risk assessments. The report avoided endorsing any particular mechanism for peer review, suggesting instead that such procedures should be tailored to meet the needs of individual agencies and should conform to certain basic procedural guidelines.[59] In general, a standing advisory committee was recommended for agencies with a large workload, whereas ad hoc panels were considered appropriate for agencies confronted with relatively few decisions. As to the composition and selection of peer-review panels, the report stressed the need for professional competence in relevant fields and for objectivity in nominating the panel members. Membership in an interest group, whether industrial or otherwise, was not viewed as an automatic disqualification. The report recommended further that review should occur before an agency publicly announced its proposed course of regulatory action. The NRC experts agreed with AIHC that the reviewing panel's determinations should not be binding, but that agencies should be obliged to discuss them and to reconsider or revise their risk assessments in accordance with the panel's suggestions.[60]

Mandatory Advice

Long before these particular proposals were placed on the policy agenda, Congress had recognized that some scientific judgments pertaining to regulation should be subjected to outside review. But neither in early legislation on scientific advice nor in more recent provisions concerning peer review has Congress endorsed the separatist approach advocated by AIHC and others in industry.

With three statutorily mandated advisory committees, the Environmental Protection Agency has clearly been a prime target of Congress's desire to inject formal scientific advice into regulatory decision-making. In 1978 an amendment to the Environmental Research, Development and Demonstration Authorization Act (ERDDA)[61] provided legislative authorization for the Science Advisory Board (SAB), an advisory body that EPA had established voluntarily in 1974. SAB is responsible for reviewing EPA's research initiatives and science-based policy determinations in every area other than pesticide regulation. The Clean Air Act Amendments of 1977 required EPA to appoint a standing committee, the Clean Air Scientific Advisory Committee (CASAC), to review the science underlying ambient air quality standards.[62] Finally, the 1975 amendments to FIFRA called upon the agency to establish a Scientific Advisory Panel (SAP) to review the scientific basis for major regulatory proposals concerning pesticides[63] and to adopt peer-review procedures for major scientific studies carried out pursuant to FIFRA.

Other major users of regulatory science, such as FDA and OSHA, are subject to notably fewer scientific review requirements than EPA. The only statutorily mandated advisory bodies for FDA are the Device Review Panels required under the medical device provisions of the Federal Food, Drug, and Cosmetic Act.[64] Congress has considered proposals to require scientific review in the assessment of food additives as well,[65] but has never enacted them into law. In practice, FDA makes extensive use of external scientific review, though the agency's five centers all have different policies with respect to obtaining expert advice. At the most formal end, the Center for Drugs and Biologics (separated into two centers following a recent reorganization) uses standing committees to review data on the safety and effectiveness of therapeutic drugs. The Center for Food Safety and Applied Nutrition, by contrast, refers to independent expert panels only on an ad hoc basis and with help from the National Toxicology Program or the Federation of American Societies for Experimental Biology.

OSHA's methods of seeking external advice are much less systematic than FDA's, perhaps reflecting OSHA's much more recent creation. The agency can in theory rely on a sister institution, the National Institute for Occupational Safety and Health (NIOSH), for technical evaluations of health and safety hazards, but for structural and other reasons (NIOSH is in HHS while OSHA is in the Department of Labor), coordination between the two agencies has not always been easy. OSHA's only standing advisory committee, the National Advisory Committee on Occupational Safety and Health (NACOSH), is responsible for guiding the agency on broad policy initiatives, such as the decision in the mid-1970s to develop a generic policy for identifying, assessing, and regulating carcinogens.[66] NACOSH has no statutory duty to perform more narrowly technical tasks, such as reviewing risk assessments, although it could presumably do so at OSHA's request.

Another agency that has been subject to mandatory peer-review requirements for some time is the Consumer Product Safety Commission (CPSC). The 1981 reauthorization of the Consumer Product Safety Act, the agency's enabling statute, provided for the establishment of Chronic Hazard Advisory Panels to review the risks of suspected carcinogens, mutagens, and teratogens before CPSC announces its intention to regulate such substances.[67] Panel members are selected with advice from the National Academy of Sciences.

Some years later, Congress responded to the political demand for peer review by writing into the 1986 Superfund amendments a review requirement for studies and research done at toxic waste disposal sites.[68] The Agency for Toxic Substances and Disease Registry must ensure that both pilot studies of health effects and full-scale epidemiological studies are peer-reviewed before they are reported or adopted. The aim of this provision clearly was to forestall disputes of the kind precipitated by Picciano's investigation of chromosome damage at Love Canal. However, the amendments did not require peer review for preliminary health assessments done by the agency to determine the need for further studies.

Taken in their entirety, the legislative provisions dealing with scientific advice display remarkably little consistency of philosophy or approach. An important indicator of congressional ambivalence is that consultation with advisory committees is legally mandated in relatively few instances. For the most part, agencies are free to structure their relationships with the external scientific community as they please, though they may confine their discretion by regulation, as FDA has done (see Chapter 8). They may choose which actions they wish to have reviewed, at what point in the course of regulatory development, and by what type of committee. Agencies may also exercise discretion in deciding how much weight they will accord to the advice or recommendations of a reviewing panel; even when review is mandated, the conclusions of the expert committee are rarely binding on the agency. Thus, the legislative framework governing the advisory process in fact affords the agencies many opportunities to adopt positions that are not expressly supported by scientists outside the regulatory establishment.

Furthermore, Congress has never developed a unified view of the structure and functions of advisory committees. Such bodies, for example, can vary enormously in their membership. Some are composed exclusively of experts from universities or independent research institutions, who are selected for their familiarity with the subject matter under consideration. Other advisory groups, however, also contain members affiliated with particular interest groups, such as the chemical industry or an environmental or labor organization. NACOSH is a multipartite committee in this sense. Reflecting their pluralistic composition, such groups are most likely to advise an agency on general policy matters. Scientists working for the government are sometimes explicitly excluded from advisory committees. Members of the Chronic Hazards Advisory Panels, for example, may not be officers or employees of the United States. But FIFRA, by contrast, authorizes peer review by scientists within or outside EPA, and scientists holding another federal office are eligible to serve on the pesticide program's Scientific Advisory Panel.

A Critical Counterpoint

The case studies presented in this chapter can be seen, at one level, as illustrating a pattern of systematic breakdowns in the management of science by regulatory agencies. They provide in this sense a powerful explanation for the emergence of scientific advice as a public policy issue in the late 1970s. As we have seen, it became the fashion in some political and scientific circles to cite these incidents as proof of official incompetence or misconduct. Conventional wisdom then dictated that controls on agency discretion should be tightened, at least where science was concerned, and proposals for regulatory peer review accordingly gained ground. There was, to be sure, less of a consensus about the way these general recommendations should be implemented by the policy system. But although the concept of peer review remained in a fluid state, both Congress and the agencies began experimenting with new institutions and procedures for legitimating the scientific underpinnings of regulatory decisions.

But was the conventional wisdom fully justified? Was there perhaps another way of reading the cases that pointed toward different conclusions, less damaging to government regulators? Of course there was. To arrive at this alternative reading, it is necessary to analyze the cases at the micro level of scientific claims and counterclaims rather than at the macro level of interest-group politics and governmental power.

Looking at the substance and context of the scientific disputes in each of these incidents, it is plausible to conclude that they had relatively little to do with the competence or incompetence of agency officials and a great deal to do with social construction, boundary work, and the politics of scientific knowledge. The central tenet of social constructivism, after all, is that perceptions of scientific "reality" are always colored by such contextual features as the scientist's professional, institutional, political, and cultural affiliations. If one accepts this view, disagreements between adversarially situated scientists—for example, an EPA consultant and the expert panels appointed to criticize his work—seem altogether predictable, even inevitable. Put differently, in a politicized environment such as the U.S. regulatory process, the deconstruction of scientific "facts" into conflicting, socially constrained interpretations seems more likely to be the norm than the exception.

The character of the scientific debates in the four cases lends at least surface support to this line of analysis. Studies in the sociology of science suggest that scientific deconstruction, when it takes place at all, centers on precisely the kinds of methodological, observational, and ad hominem issues that were raised in the controversies described above. Was the use of historical controls justifiable in the Love Canal study, and were environmental exposures properly monitored at Alsea? Did Newberne see real tumor precursors or only benign lesions? Was it possible that the "supernumerary acentric chromosomes" existed only in Picciano's imagination? Finally, were scientists like Picciano, Killian, and Schneiderman inherently untrustworthy because they (but not their critics) had permitted political ideology to interfere with their scientific methods?

The competing analyses sketched above—agency incompetence versus social construction—clearly need to be reevaluated in the light of more detailed information about the way agencies acquire, assess, and apply scientific knowledge in policymaking. Before turning to such empirical materials in later chapters, however, we must look more closely at the reasons why neither legitimation through public participation (the democratic approach) nor legitimation by outside experts (the technocratic approach) forestalled the kinds of disputes discussed in this chapter.

3

Science for the People

Judging by the volume and intensity of scholarly debate, the rise of social regulation and the resulting transformation of the American administrative process were among the defining political events of the 1970s.[1] It is surprising, in view of the level of academic interest, that one of the most interesting consequences of these developments—the evolution of policy-relevant science into a public commodity[2]—went relatively unnoticed and unremarked. Yet the cutting adrift of science bearing on policy from its traditional moorings in academic and industry laboratories emerges in retrospect as a major factor in promoting controversy over the technical underpinnings of regulatory decisions. Public debate over the legitimacy of science policy decisions intensified as both the production and analysis of scientific knowledge were increasingly drawn into public view through governmentally sponsored research, administrative rulemaking, judicial review, and, frequently, media coverage of controversies.

The push came, in the first instance, from legislative mandates to protect public health and the environment, which vastly increased the need of the executive agencies for scientific information. Congress followed through with directives for creating new federal research centers and expanding old programs, thereby enlarging the government's role in mission-oriented research. Second, the ideology of "open government" was extended to technical decisionmaking, and laws such as the Freedom of Information Act (FOIA), the Federal Advisory Committee Act (FACA), and the rulemaking provisions of major regulatory statutes affirmed the public's right of access to scientific documents and deliberations underlying policy decisions. Finally, expansive citizen suit and judicial review provisions caused a sharp rise in the volume of litigation focusing on the relationship between science and policy, and the lay judiciary emerged as an influential participant in the legitimation of science-based regulation.

A seemingly unplanned and even unforeseen consequence of making regulatory science more public was to alter the allocation of scientific authority among the major actors in the regulatory process. As regulatory agencies acquired more power to produce and interpret scientific information, the power of the courts to second-guess administrative findings also increased. Litigation became an alternative channel for debating science, undercutting the finality of both advisory committee proceedings and the agencies' own technical deliberations. To see how these shifts in the forms and locus of debate affected the use of regulatory science, we look first at the emergence of the agencies as major science centers, responsible for creating a body of public knowledge, and the accompanying growth of agency discretion in matters of scientific interpretation. We turn next to the redefinition of relations among the agencies, their scientific advisers, and the public under such laws as FOIA and FACA. Finally, we assess the impact of judicial decisionmaking on the interpretation and use of policy-relevant science. The thesis that emerges from this analysis is that the conceptual model developed by the courts to deal with technical disputes—here termed the "science policy paradigm"—failed because it created a politically unstable mix of too much administrative discretion combined with too little judicial deference.

The Rationale for Public Science

When the forerunners of the modern regulatory agencies came into being in the early twentieth century, the concept of administrative expertise was generally linked with fact-finding. In fields as diverse as taxation, transportation, communications, labor relations, and public health protection, the primary task of administrative agencies was to apply broad statutory mandates to specific facts. In order to carry out this mission, agencies had to determine what the facts were in individual cases; in this sphere of decisionmaking their expertise was usually regarded as supreme.

The post-1970 generation of administrative agencies, however, confronted technical responsibilities that were much too complex to be subsumed under the heading of fact-finding. New regulatory statutes required implementing agencies to take on a kaleidoscopic variety of scientific duties: to carry out and sponsor basic research, to conduct inspections and audits, to develop analytical and meta-analytical methodologies, and to perform risk assessments. Major modifications had to be made in the structure of the federal research establishment, and in agency processes for obtaining and evaluating scientific information, in order to accommodate these burgeoning responsibilities.

The government's suddenly expanded need for technical information provided a compelling reason for asking federal agencies to do more science. EPA, for example, faced a serious knowledge gap concerning most of the hazards it was called upon to regulate. Many toxic products and pollutants had entered commerce before there were effective controls on their manufacture and distribution, and information about them was disturbingly thin. For example, a National Research Council study concluded in 1984 that only a small fraction of the tens of thousands of commercially important chemicals had been adequately studied, and that the vast majority had not been tested at all.[3] To formulate reasonable public policies with respect to such substances, more and better information was vitally necessary. Yet the burden of producing such information could not always be rationally or fairly imposed on industry. Publicly funded research on environmental pollutants seemed particularly justified when the substance was ubiquitous (lead, asbestos) or when the causes and impacts of pollution were spread across state and national boundaries (acid rain).

Another important reason for strengthening and consolidating the government's scientific capabilities was the need to identify omissions, mistakes, and biases in data obtained from nongovernmental sources, including data demanded by law from polluters and manufacturers of toxic and hazardous substances. Agencies were called upon to specify in detail what kinds of information industry should provide in connection with applications to register new products such as drugs, pesticides, or food additives. Once the data were submitted, the quality and reliability of the information had to be checked by the receiving authority, and this task again required agencies to revamp their technical capabilities.[4]

The verification of data provided by industry involved the regulatory agencies in a second-order activity for which expertise was again a prerequisite. This was the development of guidelines, such as good laboratory practices (GLPs), which instruct manufacturers how to design, conduct, and report toxicity studies so as to ensure reliable and reproducible results. Agencies also had to develop more direct forms of quality control aimed at preventing carelessness or misrepresentation in the reporting of data. FDA, for example, began training and employing a cadre of inspectors to visit the testing laboratories that conduct health and safety studies and to audit completed studies for accuracy and consistency.

Mere increases in the scientific resources available to agencies were not the only indicator of change in the nation's expectations concerning policy-relevant science. The philosophy that government should be actively involved in the production and interpretation of knowledge permeated the majority of federal health, safety, and environmental statutes in more subtle ways. Thus, where the primary burden of generating data rests with industry (generally, in connection with the permitting or licensing of new products), agencies were expected to prescribe what information is needed, how it should be validated, and even, on occasion, by what methods it should be produced. In other cases, regulatory agencies were called upon to step in and bridge gaps in the data through their own research. Moreover, Congress implicitly designated the *interpretation* of data as a governmental function. Statutes ranging from the Clean Air Act[5] to the Toxic Substances Control Act (TSCA)[6] unambiguously stipulated that agency officials are responsible for determining from the data whether a risk exists and is worth regulating.

Public commitments to generating and interpreting science necessarily enlarged the scope of agency discretion with respect to technical issues. In spite of increased research, circumstances when risks to health or the environment can be established beyond doubt remain extraordinarily rare, and regulatory action still proceeds for the most part on the basis of indirect and uncertain evidence. Accordingly, knowledge, as Ted Greenwood and others have correctly noted, almost always has to be supplemented by discretion.

Congress generally did not see fit to constrain this type of administrative discretion through explicit statutory standards. The Delaney clause is a well-known exception: it supplies a highly restrictive regulatory formula (banning the product) once FDA finds that the evidence points to a risk of cancer. The legislature determined in this specific instance that a particular type of evidence, "appropriate" cancer studies, should automatically constitute grounds for regulation. FDA has little discretion to deviate from this preordained policy, though it does have some leeway to determine whether or not the scientific evidence is "appropriate" to trigger action.[7] In most cases, however, the agencies not only enjoy greater freedom to decide whether a regulatable risk exists but may look to nonscientific evidence in making this decision. For example, statutes such as TSCA and the Federal Insecticide, Fungicide, and Rodenticide Act[8] require agencies to show that a risk is "unreasonable" before undertaking regulatory action. Here, the trigger for regulation is a composite determination that takes into account both the nature and magnitude of the risk and the social consequences of regulation.[9]

The "New" Expert Agency

Congress's commitment to creating a public scientific base for public policy also manifested itself in the proliferation of new federal research programs and in the reorganization of existing research activities. Thus, the National Institute for Occupational Safety and Health (NIOSH) was created in 1970 to advise the Department of Labor on the scientific and technical aspects of worker protection. The 1971 National Cancer Act elevated the organizational status of the National Cancer Institute (NCI) within the National Institutes of Health (NIH) and expanded its research capabilities. The National Institute for Environmental Health Sciences (NIEHS) was established in 1969 as a nonstatutory entity within NIH; its relatively rapid growth during the next several years reflected a mounting national concern with problems of pollution-related ill health.[10] A U.S. Army biological research facility was converted in 1971 into the National Center for Toxicological Research (NCTR), currently administered by FDA. In 1978 the toxicological research programs of NIEHS, NIOSH and NCTR were consolidated under the National Toxicology Program (NTP), of which NCI is also a participant. With a budget of $77.4 million in fiscal year 1984,[11] NTP emerged as the primary institutional focus of federal efforts to forge stronger links between toxicological research and regulatory needs.

The most dramatic extension of public scientific activity, however, occurred at the Environmental Protection Agency, which was established by a presidential reorganization plan in 1970.[12] Within a few years, EPA emerged as the epitome of the "new" expert agency, with far-flung research enterprises dedicated to environmental protection and risk reduction. The agency's in-house research capabilities were distributed among more than a dozen laboratories focusing on environmental engineering, health effects, environmental impacts, and environmental monitoring. These laboratories are dispersed across several states, including Georgia, Nevada, North Carolina, Ohio, and Oregon. In 1983, the agency's total scientific and engineering staff consisted of 3,841 permanent, full-time employees,[13] and its R&D budget for 1986 amounted to $320.5 million, 70 percent of which was earmarked for extramural research.

EPA's five research committees—responsible for water, air and radiation, hazardous wastes, multimedia energy, pesticides and toxics—delineate the areas of greatest scientific uncertainty and design research programs for reducing them. High priorities on EPA's research agenda for 1986–1991 included studies of acid deposition, hazardous wastes, biotechnology, reproductive toxicology, and exposure assessment.[14] In addition, about 15 percent of EPA's research budget is devoted to exploratory research on emerging environmental issues; by 1986, EPA had established eight exploratory research centers at academic institutions across the country, extending the capabilities of its own laboratories and linking the agency more firmly to nongovernmental scientific and technical communities.

EPA's obligations in the scientific domain also include, increasingly, the development of analytical methods for interpreting uncertainty. The agency's expertise in such areas as toxicology, epidemiology, and statistics has grown, in keeping with the need for more formal and systematic analyses of risk. A 1985 EPA publication stated that some three thousand documents related to risk assessment were produced throughout the agency in the course of a single year.[15] The Office of Research and Development (ORD) is responsible for working with the various program offices to ensure that these efforts are consistent and technically competent.

While EPA's research programs outstrip those of any other federal regulatory agency in scope and diversity, several of EPA's sister agencies have also played a significant part in building up the public knowledge base for use in health, safety, and environmental decision-making. Of these, the Food and Drug Administration is the most securely established. Descended from an agency formed to implement the 1906 Pure Food and Drugs Act, FDA can claim a continuous involvement in risk regulation and science policy for more than eighty years. In 1983, the agency's scientific and engineering staff numbered 3,185 out of a total workforce of 6,865.[16]

FDA's organization and research commitments have changed over time to reflect increases in the agency's regulatory burden and its associated need for in-house research and expertise. An earlier bureau structure, built around specific product classifications, was replaced by a number of centers with more rationally distributed obligations. More specifically, in 1982, FDA Commissioner Arthur Hull Hayes combined the Bureau of Drugs and Bureau of Biologics into the Center for Drugs and Biologics, apparently hoping that drug regulation would benefit from the longstanding research strength of the Bureau of Biologics.[17] But a reorganization in 1987 again split up these functions, creating two new centers for Drug Evaluation and Research and Biologics Evaluation and Research. As their titles indicate, a cardinal objective of the new centers was to give greater prominence to research in the regulatory programs for both drugs and biologics.

Through NCTR, FDA participates in the national program to increase toxicological information pertinent to public health. Although NCTR, like most federal agencies, suffered serious budget cuts in the early 1980s, its publication record and acceptance rate on manuscripts both improved markedly during this period, suggesting an overall increase in the productivity of its basic research.[18] Besides sponsoring basic research at NCTR, FDA has played a leading role in shaping federal science policy on a variety of controversial issues, such as risk assessment, good laboratory practices,[19] and the use of advisory committees.

By comparison with EPA and FDA, smaller regulatory agencies such as the Consumer Product Safety Commission make only minor independent contributions to policy-relevant R&D. For example, in fiscal year 1984 CPSC spent $1.7 million in cancer-related activities, whereas FDA spent $16 million and EPA $11.4 million.[20] But in the aggregate these lesser agencies also add substantially to the store of expertise available to governmental decisionmakers. More important, their activities help substantiate the proposition that the federal government is not only responsible for regulating technological hazards but, where necessary, for carrying out the science needed to make such decisions.

Scientific Advice and Open Government

While reaffirming the need for more publicly generated scientific information, Congress also enacted laws to enhance the public's watchdog role over the process by which agencies seek scientific advice. In a marked swing toward democratic decisionmaking, both the Federal

Advisory Committee Act and the Freedom of Information Act asserted, in effect, that no component of regulatory decisionmaking was too arcane or technical to be entirely isolated from public review and criticism.

The Federal Advisory Committee Act

The enactment of FACA in 1972 responded to a growing concern among legislators and the public that the unchecked proliferation of advisory committees and their closed methods of operation were creating a "fifth arm of government" standing outside the networks of democratic control.[21] FACA was to do for some 5,000 federal advisory committees what the Administrative Procedure Act had done for the administrative agencies a generation earlier. It was to ensure, in the first place, that uniform standards and procedures would govern the creation, organization, and functioning of advisory committees, and that ineffective or duplicative bodies would be eliminated. Secondly, Congress wanted to open up public access to committee operations and records, thereby reducing the possibility of bias in the formulation of expert recommendations.

FACA prohibits the establishment of any advisory committee unless it is specifically authorized by statute, by the president, or by an agency head who has determined, as a matter of formal record, that the committee is in the public interest. Advisory committees are prohibited from meeting or taking action until a charter has been filed with the director or agency head and with appropriate standing committees in Congress. In addition, notice of the creation of an advisory committee must be given in the *Federal Register* so that the public has an opportunity to challenge its establishment. Other significant requirements include provisions for open meetings, detailed transcripts, limited rights of public participation, and the attendance of a federal government representative at advisory committee meetings.

FACA has been construed as applying only to groups with an established structure and defined purpose. Thus, biweekly White House meetings with business organizations and other private-sector groups were not considered subject to FACA because they were unstructured, informal, and not conducted for the purpose of obtaining advice on specific subjects indicated in advance.[22] This exception is probably immaterial to most scientific advisory committees, since these are usually constituted for specific purposes and often are invited to answer questions posed in advance by the sponsoring agency. But FACA also does not apply to persons or organizations having a contractual relationship with the agencies. This exception might cover the National Academy of Sciences,[23] as well as independent consulting firms hired to provide technical services to regulatory agencies.[24] In addition, the Act does not apply to advice that was neither explicitly nor implicitly solicited by the receiving official.

To assure the objectivity of advisory committee decisions, FACA requires that committees be "fairly balanced in terms of points of view represented and functions to be performed."[25] The purpose of this constraint is to prevent agencies from loading the dice in favor of particular analytical approaches or policy outcomes. While the law provides no explicit standards by which to evaluate fair balance, a federal district court has held that the provision should be construed "in terms of. . . the functions to be performed by the advisory committee."[26] This reading suggests that "balance" must be defined in relation to the substance of the committee's deliberations. Where the issues are broad and policy-oriented, advisory committees should also be broadly representative, along technical, social, and political dimensions. By contrast, where the issues are more technical, as in the area of risk assessment, FACA may be satisfied by an adequate representation of relevant scientific viewpoints.[27]

FACA authorizes private lawsuits to enforce its provisions, but a citizen's right to challenge agency action under the Act may be limited by standing requirements. A citizen must first show a specific injury flowing from the alleged violation. The D.C. Circuit has held, for instance, that neither an individual consumer nor a U.S. senator had standing to complain that the National Petroleum Council was not fairly balanced in membership, since the plaintiffs' injuries were "speculative and conjectural in the purest sense."[28] Other FACA plaintiffs, however, have had more success. Several courts have found that persons denied admission to improperly closed committee meetings had standing because their rights to access were curtailed.[29] In addition, the D.C. Circuit has indicated that a person may have standing to challenge the balanced membership requirement either when denied representation on the advisory committee[30] or when that person has "a direct interest in the committee's purpose."[31]

A frequently noted weakness in the Act is that it provides no clear remedies for noncompliance, leaving courts at liberty to design their own remedies. If a suit is begun before a violation of FACA occurs, the offending behavior may be enjoined.[32] After a violation, courts could in theory fashion an exclusionary rule barring agencies from using improperly obtained advice.[33] In an era of judicial restraint, however, courts are unlikely to announce such a rule. A more likely response would be to find that a procedural error in using advisory committees was not serious enough to justify overruling an agency decision based on a complex technical record.[34]

The Freedom of Information Act

Focusing on the written rather than the spoken record of decision-making, the Freedom of Information Act of 1970 established a statutory presumption that the public should have access to government documents except where such material is specifically exempted by law. Congress later incorporated the FOIA exemptions into both FACA and the Sunshine Act, because they represent a comprehensive and deliberate balancing of the values of public disclosure against the need to protect information. Potentially most significant from the standpoint of expert advice is exemption five of FOIA,[35] which covers intra-agency memoranda that would not be discoverable by a party in litigation with the agency. Some of the most important legal controversies arising under FACA immediately after its enactment concerned the extent of exemption five's applicability to advisory committees. In interpreting the exemption within the context of FACA, the Office of Management and Budget (OMB) at first indicated that meetings or portions of meetings could be closed to the public not merely when an exempt document was under consideration but also when a committee was exchanging opinions that would, if written, fall under exemption five. A number of judicial decisions, however, have construed exemption five to allow greater public access than was contemplated by OMB.[36]

If an advisory committee is not regarded as an "agency" under FOIA, its work products arguably cannot be covered by an exemption relating to "inter-agency or intra-agency memorandums." The federal district court for the District of Columbia applied this reasoning to require disclosure of meeting transcripts in a case involving FDA's Over-the-Counter Antacid Drugs Advisory Review Panel.[37] The court concluded, moreover, that exemption five cannot be used to withhold transcripts of advisory committees where such records are not part of the parent agency's deliberative process. Such broad application of exemption five would, in the court's view, have jeopardized FACA's basic objective of liberalizing public access to advisory committee meetings. Other FOIA provisions that could potentially restrict the openness of advisory committee meetings include exemption four,[38] which prohibits disclosure of trade secrets and commercial or financial information, and exemption seven, covering investigatory files compiled for law enforcement purposes.

Judicial Review of Science Policy

The legal and institutional developments described above attest to Congress's active involvement in laying the infrastructure for governmental and citizen participation in policy-relevant science. But from the standpoint of scientific decisionmaking, the emergence of the federal courts as aggressive participants in the regulatory process proved even more consequential. Judicial intervention added a distinctively populist cast to scientific disputes, both because the courts provided a forum for citizen complaints regarding regulatory science and because lay judges, with no relevant technical training, became the final arbiters of such disputes.

Beginning in the early 1970s, judges responded with alacrity to charges by citizen intervenors that the agencies were acting too slowly in the face of strong mandates from Congress and accumulating evidence of risk. In order to spur the regulators to action, the courts proved more than willing to steep themselves in the intricacies of environmental and public health regulation and to make their own independent evaluations of the agencies' technical determinations. But although such activism unquestionably stimulated action, particularly at EPA, its impact on the quality of scientific decisionmaking was more equivocal. Judicial review produced a paradigm for resolving science policy disputes, but one that appears in hindsight to have been founded on an overly simplified view of both science and policy.

The Science Policy Paradigm

By committing greater responsibility for scientific interpretation to the agencies, Congress, as we have seen, entrusted to administrators a new kind of hybrid decisionmaking—"science policy"—encompassing both scientific and policy considerations.[39] Given their unique combination of technical and policy skills, the new expert agencies appeared, in principle, well qualified to make science policy determinations. But the full implications of agency authority in this area began to take shape only gradually, as courts construed new statutory language aimed at health, safety, and environmental protection. It was ultimately the courts, rather than Congress, that articulated the basic propositions of decisionmaking that constitute what I call the "science policy paradigm."

The paradigm comprises three key elements, each of which deeply influenced agency procedures for evaluating science and the structure of the scientific advisory process. The first is the notion that agencies should be permitted to make regulatory decisions even on the basis of imperfect knowledge (that is, suggestive rather than conclusive evidence).[40] The second element, a corollary of the first, is that a science policy determination may be regarded as valid even if the scientific community does not universally accept it as such.[41] A third and closely related principle is that when experts disagree about the validity or interpretation of relevant data, the administrative agency should have the authority to resolve the dispute consistently with its overall legal mandate. Put another way, when science alone is incapable of providing unique answers to questions about risk, the choice among conflicting answers should be made by the politically accountable agency in accordance with its lawful regulatory mission.[42]

These views were perhaps most persuasively set forth in *Ethyl Corp. v. EPA*, the earliest in-depth exploration of science policy by a reviewing court. Judge Skelly Wright of the District of Columbia Court of Appeals observed that in implementing a precautionary statute, such as Sec. 211(c)(1)(A) of the Clean Air Act, an agency administrator could appropriately act on the basis of less than perfect knowledge; specifically, conclusions could be drawn from "suspected, but not completely substantiated, relationships between facts, from trends among facts, from theoretical projections from imperfect data, from probative preliminary data not yet certifiable as 'fact,' and the like."[43] Though the opinion did not explain just what was meant by "imperfect data" or "data not yet certifiable as 'fact'," such language seemed to authorize the agencies to accept science not yet endorsed by the scientific community. At the very least, *Ethyl* appeared to give administrators considerable discretion to decide how they should resolve disputes over scientific facts, theories, and methodologies.

Further developing these themes, the court rejected the notion that "facts" relied on by agencies must conform to scientific standards of certainty:

> Petitioners demand sole reliance on *scientific* facts, on evidence that reputable scientific techniques certify as certain. Typically, a scientist will not so certify evidence unless the probability of error, by standard statistical measurement, is less than 5%. That is, scientific fact is at least 95% certain.
>
> . . .
>
> Agencies are not limited to scientific fact, to 95% certainties. Rather, they have at least the same fact-finding powers as a jury, particularly when, as here, they are engaged in rule-making.[44]

Subsequent cases held that in regulating carcinogens, a health hazard that courts regarded as particularly "sensitive and fright-laden,"[45] it was appropriate to uphold "action based upon lower standards of proof than otherwise applicable."[46] *Dow Chemical v. Blum*[47] provided a striking example of judicial acceptance of such a lowered quantum of proof. The case involved a challenge to EPA's emergency suspension of the herbicide 2,4,5-T, but predated the rejection of the Alsea II study by EPA's advisory committee and the scientific community. Dow attacked both the study's design and use, including EPA's data collection procedures, statistical interpretations of the data, and assumptions regarding levels of human exposure to the compound. The court considered these arguments substantial, but conceded that EPA's conclusions could not be set aside unless they were completely without foundation or the agency had made a clear error of judgment. Since the record showed some support for the agency's position on all of the points raised by Dow, the court decided "with great reluctance" not to stay the emergency suspension order.

In order to expedite action, the D.C. Circuit was even prepared to relax the procedural standards conventionally applied to regulatory decisionmaking. In *Hercules, Inc. v. EPA,* for example, the court alluded to "the rule of ancient origin that expedition in protecting the public health justifies less elaborate procedure than may be required in other contexts."[48] Such observations underscore the fact that courts in this period were not interested in subjecting the scientific basis for policy to exceptionally rigorous screening by scientific supervisors. The concept of scientific peer review seemed far removed from the judiciary's thinking.

Judicial concern about subjecting agency decisions to external criticism focused instead on review by the general public. The U.S. Court of Appeals for the First Circuit noted, for instance, in a case involving the Seabrook nuclear power plant, that consulting a scientific advisory committee was no substitute for rulemaking accompanied by full public participation.[49] During the licensing proceedings, EPA had to issue a permit for the plant's cooling system pursuant to the Federal Water Pollution Control Act. The regional administrator's initial decision to deny the permit was reversed by the EPA administrator, partly on the basis of information and recommendations supplied by a specially appointed scientific panel. The Seacoast Anti-Pollution League, an activist citizen group, challenged the adequacy of this closed-door procedure to fill the alleged gaps in the record. The court agreed with the petitioners that the administrator should either have based his decision on the existing record or held hearings permitting interested parties to cross-examine panel members.

The presumptions of the science policy paradigm were further refined in cases holding that decisions should not be reversed simply because some scientific authorities contradicted the agency's findings.[50] Indeed, the *Hercules* court took pains to affirm that where methodological issues are contested it is the expert agency's function to select the appropriate analytical approach:

> The choice of a suitable technique for estimating toxicity risks is a decision "on the frontiers of scientific knowledge." The relative scientific and administrative advantages of different techniques are incapable of precise comparison by this court. Decision between the alternatives is a quintessential policy judgment within the discretion of EPA. We cannot accept Hercules' notion that the administrator of the agency created to protect the environment lacked even the *capability* to exercise the discretion with which he was entrusted by Congress.[51]

One early environmental decision that cut in a somewhat different direction was *International Harvester Co. v. Ruckelshaus*,[52] where the D.C. Circuit invalidated an EPA decision for failure to provide adequate scientific justification. The court was especially troubled by the agency's unexplained divergence from the methods employed by an NAS panel to determine the feasibility of the 1975 auto emission standards prescribed under the Clean Air Act. The NAS report concluded that technology was not available to meet these standards in time. EPA's approach, by contrast, suggested that the standards were feasible and this led the agency to deny the automobile industry a one-year stay in achieving compliance. The court acknowledged that the Academy panel's conclusion was a "necessary but not sufficient condition" for suspending the standards and that EPA was "not necessarily bound by NAS's approach, particularly as to matters interlaced with policy and legal aspects."[53] Nevertheless, the court felt that EPA was not free to substitute its own scientific assumptions for those used by the Academy without explicitly justifying those choices, perhaps by adducing new research and experience to support its position.

The court's refusal to defer to EPA in *International Harvester* can be explained on grounds peculiar to that decision. The case arose in EPA's infancy and involved the court for almost the first time in evaluating the agency's use of predictive models. Moreover, the NAS study was specifically mandated by Congress, apparently to build a technical rationale for delaying a costly and perhaps needlessly draconian regulatory intervention. Under the circumstances, EPA's novel and untested methodology was bound to seem not just unscientific but misguided. Indeed, William Pedersen, a former EPA lawyer, concluded that *International Harvester*'s real significance was to warn EPA against regulatory tunnel vision: the decision was a valuable reminder that eagerness to act is no excuse for overlooking methodological uncertainties or for underestimating the costs of too much regulation.[54] In this respect, the case provided a foretaste of things to come.

Discretion and Deference

The science policy paradigm contemplated, in theory, that high administrative discretion to resolve scientific disputes would be coupled with equally high judicial deference. But as courts began reviewing increasing numbers of science-based regulatory decisions in the 1970s, the theory of deference gave place to a widespread practice of judicial intervention. The instability of the resulting high-discretion, low-deference approach to making science policy became a major factor in the eventual weakening of the paradigm.

Under the Administrative Procedure Act (APA), as well as under the review provisions of individual regulatory statutes, American courts have a duty to monitor the substantive rationale for regulatory decisions and to determine whether the findings made by administrators are in fact adequate to support their decisions. Proposed regulatory decisions that are not sufficiently anchored in the record of decision-making may be overruled either because they are "arbitrary, capricious, [or] an abuse of discretion"[55] or because they are not supported by "substantial evidence."[56] The practical distinction between these two legal formulations is debatable, and federal courts are now generally agreed that in reviewing decisions based on a mixture of scientific and nonscientific considerations the two standards are essentially interchangeable.[57] In deciding how thoroughly a court should scrutinize a science policy decision, it is therefore immaterial whether the statute calls for "arbitrary and capricious" or "substantial evidence" review.[58]

Focusing instead on the purposes that judicial review is intended to serve, courts have come to the conclusion that substantive review, including review of an agency's scientific determinations, is the area where they should step most lightly.[59] Not only are these matters explicitly delegated by Congress to expert agencies, but judges have no independent basis for second-guessing administrators on such highly technical issues. Judicial opinions on this subject are shot through with professions of reluctance to substitute the scientific judgments of courts for those of agencies. Yet judicial practice, especially in the early years of environmental decisionmaking, frequently fell out of step with the rhetoric of restraint.

A turning point in the development of modern environmental law was the declaration by federal judges that it was their duty to force agencies to take a "hard look" at the technical evidence before them.[60] While courts expressed no desire to make scientific findings on their own, they insisted on holding the agencies accountable for a full and reasoned explanation of their technical determinations. According to an authoritative opinion of the D.C. Circuit Court of Appeals, it was entirely appropriate for a reviewing court

> first, to insist upon an explanation of the facts and policy concerns relied on by the Agency in making its decision; second, to see if those facts have some basis in the record; and finally, to decide whether those facts and those legislative considerations by themselves could lead a reasonable person to make the judgment that the Agency has made.[61]

In order to ensure that the agency had acted "reasonably," the reviewing judge was entitled by this standard to probe rather deeply into the administrator's scientific thought processes.

The "hard look" metaphor in this way subtly became a rationale for letting the courts themselves look hard at the scientific arguments underlying agency decisions; in other words, it provided a vehicle for remarkably intrusive review in just those areas where deference was in principle most warranted.[62] In a growing body of cases, particularly those involving decisions not to regulate, the courts set aside decisions because they found the agency's scientific record or reasoning inadequate.

One strategy the courts developed to bridge the talk of restraint and the practice of intervention was to ask the agencies to follow more stringent rulemaking procedures than they were required to do by law.[63] It became a prominent theme of the D.C. Circuit Court in the 1970s that judges, being "technically illiterate,"[64] should limit themselves to ensuring that agency officials had followed proper procedures in examining scientific evidence. Chief Judge David Bazelon, a leading proponent of this view, insisted that careful monitoring of agency procedures would not only encourage more care and completeness in the analysis of science but would also promote responsible peer review. The court's duty, Bazelon argued, was to get decisionmakers to "disclose where and why the experts disagree as well as where they concur, and where the information is sketchy as well as complete"; other experts steeped in the subject matter could then "evaluate the agency's factual determinations, bring new data to light, or challenge gaps in reasoning."[65]

To the extent that the D.C. Circuit expressed a preference for particular procedures, its primary emphasis was on adversarial approaches, especially cross-examination. Judge Bazelon often asserted that a clash of opposing expert viewpoints was indispensable for developing an adequate record for judicial review.[66] In *Natural Resources Defense Council v. Nuclear Regulatory Commission*,[67] he listed a variety of procedures that agencies might adopt to encourage a "genuine dialogue" on technical issues. Among them was a suggestion that agencies should make use of "technical advisory committees comprised of outside experts with differing perspectives."[68] Bazelon did not specify whether the "perspectives" in question should be disciplinary or political, but he seemed quite certain that the confrontational procedures of the courtroom were ultimately best suited to exploring science.[69]

Recent research on regulatory decisionmaking suggests that Bazelon's confidence in adversarial procedures as a way of illuminating scientific disputes was probably misplaced. Adversarial proceedings, it is now recognized, can polarize opinion, distort evidence, and multiply the grounds for disagreement. Moreover, they place the ultimate decisionmaker in the awkward position of having openly to take sides in a scientific debate, thus increasing his vulnerability to challenge.[70] EPA's experimental initiatives with the concept of negotiated rulemaking in the 1980s provide one index of the regulatory system's disenchantment with confrontational proceedings.[71]

Bazelon's thinking was problematic from a jurisprudential perspective as well, for it gave courts an almost unrestricted license to impose new procedural requirements on agencies whenever they were dissatisfied with the quality of an administrator's scientific information, assumptions, or reasoning. Administrative lawyers were not wholly surprised, therefore, when the Supreme Court firmly rejected Bazelon's approach in *Vermont Yankee Nuclear Power Corp. v. Natural Resources Defense Council.*[72] In an unusually strong, unanimous opinion,[73] the Court held that judges could not require an agency to adopt any procedures except those specifically authorized by Congress or needed for due process. On its face, the case seemed to rule out the possibility of invalidating administrative decisions simply because the agency did not allow cross-examination, solicit divergent expert opinions, or submit its science policy determinations to peer review.

But *Vermont Yankee* did not reject the hard look doctrine. On the contrary, Justice Rehnquist noted that courts might continue to remand regulatory decisions for deficiencies in the record and that such remands, in turn, might lead the agency to adopt procedures over and above those originally employed in developing the record: "If the agency is compelled to support the rule which it ultimately adopts with the type of record produced only after a full adjudicatory hearing, it simply will have no choice but to conduct a full adjudicatory hearing prior to promulgating every rule."[74] It was not until *Baltimore Gas and Electric Co. v. Natural Resources Defense Council,*[75] that the Supreme Court definitively reasserted the principle of judicial deference on technical issues, stating that a court "must generally be at its most deferential"[76] when reviewing determinations at the frontiers of science.

A return to judicial deference also explains the D.C. Circuit's interesting opinion on *ex parte* communications in *Sierra Club v. Costle.*[77] The case arose from a complaint by the Environmental Defense Fund

(EDF) that EPA's new source performance standards for coal-burning power plants were impermissibly tainted by an "ex parte blitz" from the coal industry. EDF objected to EPA's decision to continue receiving comments on the proposed limits on sulfur dioxide emission long after the close of the official comment period. EDF also challenged as improper a number of meetings held by EPA with industry representatives, White House officials, and members of Congress during the postcomment period. In a balanced and detailed opinion, Judge Wald concluded that neither of these reasons provided a sufficient basis for overruling EPA. None of the documents vital to the agency's support for the rule had been submitted so late as to preclude effective public comment.[78] Further, the court was persuaded that informal communications among the agency, the regulated industry, and government officials—far from constituting impermissible political influence—were actually important to the functioning of a legitimate regulatory program.[79] A blanket ban on meetings with individuals outside EPA was therefore unwarranted.

In *Sierra Club v. Costle,* the D.C. Circuit in effect resisted the temptation to hold EPA to a high standard of technical rationality, even though the regulation under review rested to a large extent on scientific and technical considerations. As Judge Wald noted, "we do not believe that Congress intended that the courts convert informal rulemaking into a rarified technocratic process, unaffected by political considerations."[80] Implicit in this statement was a criticism of the earlier activist stance of the court, whereby insistence on procedural regularity had forced inherently political questions to be decided at the level of technical debate. Political influences, the court suggested, would only be judged improper if they caused the agency to consider factors not made relevant by its governing statute.[81] In the oscillation between technocratic and political conceptions of the agencies' role, the *Sierra Club* opinion clearly came down on the political side.

The Weakening of the Paradigm

The retreat from judicial activism in the foregoing cases heralded, in effect, a return to the "official" version of the science policy paradigm: a high (administrative)–discretion, high (judicial)–deference approach to making science policy. Other cases, however, indicated that courts were beginning to have second thoughts about the implications of high administrative discretion in the resolution of technical controversies. In *Industrial Union Department v. American Petroleum Institute,*[82] the Supreme Court expressed serious dissatisfaction with the paradigm's third major element: that it is within the administrator's uncontrolled discretion to choose among competing expert opinions. Asked to review OSHA's proposed occupational standard for benzene, a plurality of the justices held that OSHA could not promulgate a more stringent standard unless it could show, on the basis of substantial evidence, that the existing exposure limit posed a significant risk to worker health. Most interestingly from the standpoint of science policy, the court indicated that OSHA should have used mathematical models to calculate the level of risk.[83] This reversed the agency's own determination that quantitative risk assessment was not required in regulating benzene, a known carcinogen.[84] The practical effect of the decision was to overturn a regulatory agency's expert judgment on a methodological issue, precisely the kind of determination committed to agency discretion pursuant to the science policy paradigm.

A similar retrenchment occurred in *Gulf South Insulation v. Consumer Product Safety Commission.*[85] Here the U.S. Court of Appeals for the Fifth Circuit invalidated CPSC's attempt to ban urea-formaldehyde foam insulation (UFFI) on the basis of animal studies showing that formaldehyde gas causes cancer. The court found fault with the agency's exposure measurement techniques and, more important, with its decision to estimate the risk of cancer to humans from a single animal study. In the court's view, the uncertainties in the data base were too large to permit the use of "an exacting, precise and extremely complicated risk assessment model."[86] The decision was widely criticized on the grounds that the court went too far in its reanalysis of CPSC's scientific determinations.[87] For present purposes, however, the notable feature of both the benzene and UFFI cases was the failure of an administrative agency to persuade a reviewing court to accept its claims of expertise as controlling.

In the cautious regulatory environment ushered in by the Reagan administration, courts would arguably be most deferential to agency decisions if the supporting scientific data and analyses (in the form of a risk assessment, criteria document, or position paper) were routinely peer-reviewed. Prior scrutiny by nonagency scientists might satisfy the courts that the agency's resolution of disputed technical evidence was backed by more than mere claims of administrative expertise. But the only significant case to date on the necessity for regulatory peer review delivered a mixed message. In *API v. Costle,*[88] the D.C. Circuit sustained EPA's primary and secondary ambient air quality standards (NAAQS) for ozone in spite of serious deficiencies in the agency's interactions with its Science Advisory Board (SAB). Consultation with SAB was mandated by Congress, but EPA had only partially fulfilled its statutory obligation to seek the Board's advice on its criteria document and final standard.[89] Although the court was troubled by these procedural errors, it ultimately deferred to EPA, thereby reaffirming the spirit of *Vermont Yankee* that judges should be extremely reluctant to overturn an agency's substantive findings on purely procedural grounds.

Inconsistent decisions and wavering judicial support underscored the political fragility of the science policy paradigm and added weight to industry's demands for better quality controls on regulatory science. The chemical industry, for example, had never conceded that agencies should have as much discretion to make risk-assessment determinations as they did pursuant to the classic version of the paradigm. Indeed, controversies over substances such as benzene and formaldehyde represented, at bottom, a tug-of-war between regulators and public interest groups on the one hand and chemical producers on the other about the extent to which the science policy paradigm could validly be used to resolve disputes resulting from scientific uncertainty. Industrial opposition to particular regulatory outcomes quickly developed into demands for a return to more technocratic processes for considering regulatory science. But although the message "leave science to the scientists" was superficially appealing, it failed to address the underlying problem of defining what counts as "science" in areas of methodological uncertainty and political conflict.

In arguing against discretion and for peer review, industry representatives resorted to boundary work designed to shift the line of demarcation established by the courts between "science" and "policy." Risk-assessment guidelines, for instance, were generally characterized as science, suitable for resolution by accredited expert bodies like the National Academy of Sciences. Industrial groups were convinced that these technocratic organizations would reach conclusions that were scientifically more conservative, hence more sympathetic to business interests, than those advocated by administrative agencies or the courts. The boundaries drawn by industry, however, were contested to varying degrees and in different ways by actors pursuing other social and political agendas. The struggle for control over regulatory policy was thus played out in part on the fields of discourse, as terms like "science," "policy," and even "peer review" were redefined to fit different conceptions of the relationship between science and power.[90] As the next chapter argues, conflicts such as these were only to be expected in light of the characteristics of regulatory science and its political context.

4

Peer Review and Regulatory Science

Would-be reformers, especially those disenchanted with the legal system, have understandably looked to peer review as a promising option for restoring to scientists stricter control over regulatory science. For more than three hundred years, Western science has relied on peer review as the primary means of identifying work that deserves to enter the domain of certified knowledge. Refereeing procedures have come to be regarded as the most effective method of validating science in two quite different spheres of professional activity: prepublication review of journal articles and screening of applications by federal research-sponsoring agencies. There is thus an appealing logic to the syllogism that links peer review to "good science" in the regulatory process. Peer review, so the argument runs, has clearly been successful in screening out good science from bad in the research context. Regulatory agencies also need a technique for separating good from bad science. Therefore, wider adoption of peer review by regulators will assure them access to better science.

But does this argument stand up to critical analysis? The validity of the first premise, in particular, has been challenged in recent years by a growing body of investigations questioning the reliability of conventional peer review. These studies point to two sets of conclusions, each with disturbing implications for regulatory peer review and, more broadly, for the model of technocratic legitimation in regulatory science. Recurrent cases of plagiarism, fraud, and misconduct have demonstrated, in the first instance, that peer review is by no means a fail-safe method of assuring the quality and integrity of scientific claims. Second, and perhaps more significant, studies of peer review have provided a window on the problematic character of validity in science. Advocates of peer review tend to assume that the line between acceptable and unacceptable science is objectively verifiable, available to be discovered by any competent reviewer. Yet the empirical literature on peer review suggests almost the opposite: that standards for deciding what is acceptable are matters of negotiation and compromise, and that peer review is simply part of the process of construction by which scientists certify some claims and conventions as valid. Seen in this light, peer review works best when there is already substantial agreement among researchers about theories, experimental methods, and the goals and objectives of measurement, a situation that obtains at best unevenly across the many subdisciplines of science.

These findings suggest that the benefits of peer review (and, more generally, of scientific advice) for regulatory science may have been oversold. If peer review is most effective at cementing consensus among scientists of similar disciplinary training and outlook, then the chances are that the process will prove much less successful when it is carried out under uncertain theoretical and methodological constraints, as well as under pressure of politics and regulatory deadlines. To explore these implications more closely, this chapter first looks at some of the current debates surrounding peer review in the scientific and social science literature. It then examines the contextual differences between regulatory science and research science and asks how these differences are likely to affect the integration of scientific advisory and review processes into regulatory decisionmaking.

The Traditions of Peer Review

Peer review has been described as a mirror of science.[1] To understand why peer review works, one must therefore turn to theories of why science itself works. Robert K. Merton, America's pioneering sociologist of science, provides a natural starting point.

In 1942, troubled by what he perceived as dangerous attacks on the integrity of science, Merton set out to do for science what Milton did for God in *Paradise Lost:* "to vindicate the ways of science to man."[2] In a classic essay entitled "The Normative Structure of Science," Merton tried to elucidate the social conditions and institutional arrangements that account for the flourishing of modern science. Scientific activity derives its authoritative status, he suggested, from a set of shared norms, or institutional imperatives, that appear to guide the conduct of research: universalism, communism (also called communalism), disinterestedness, and organized skepticism. The principle of universalism ensures that scientific claims will be evaluated on the basis of preestablished, impersonal criteria, without regard to the personal or social attributes of the scientist. Communism is the norm that prevents exclusive ownership of scientific findings by any individual or group. Disinterestedness or the absence of personal motivation in research is a norm that the scientific community achieves, in Merton's view, through vigorous self-policing. Finally, organized skepticism is the shared intellectual attitude among professional scientists that ensures detached scrutiny of beliefs and claims in terms of empirical and logical criteria.

The norms of science appear at first glance to provide a satisfying explanation for the successful performance of peer review. If scientists adhere to the Mertonian norms in their research, particularly disinterestedness and organized skepticism, we should expect these norms to govern the process of peer review as well, leading to high professional and ethical standards in scientific refereeing. Much controversy, however, has swirled around Merton's exposition of the norms, and it is now widely recognized that what he described was at best a set of conditions necessary for the success of science. As an account of actual scientific behavior, the norms are wholly inadequate. They presuppose a kind of purity in the conduct of research that is belied by the findings of the social constructivists. Their work suggests that the norms in reality function as a potent resource for science. By claiming adherence to these idealized rules, in peer review as well as in research, science projects to the world not what it actually is but what it aspires to be and, above all, how it would like to be seen.[3]

To explain the success of peer review, then, we must turn to other features of the scientific enterprise, and Merton's writings again point the way. In describing the British Royal Society's adoption of a referee system in the late seventeenth century for its official publication, *Philosophical Transactions*,[4] Merton and Harriet Zuckerman called attention to the changes within science that made such a system indispensable:

> Ingredients of the referee system were thus emerging in response to distinctive concerns of scientists taken distributively and collectively. In their capacity as producers of science, individual scientists were concerned with having their work recognized through publication in forms valued by other members in the emerging scientific community who were significant to them. In their capacity as consumers of science, they were concerned with having the work produced by others competently assessed so that they could count on its authen-

> ticity. In providing the organizational machinery to meet these concerns, the Royal Society was concerned with having its authoritativeness sustained by arranging for reliable and competent assessments.[5]

The institutionalization of peer review, according to this account, was facilitated by the emergence of a professional scientific community concerned with upholding the interests of its members in recognition, authority, and above all, dependable knowledge. Peer review, in other words, was a social compact created and sustained by the self-centered communal needs of modern science.

Linking peer review to professionalism in this way helps explain why the procedure commands wide allegiance, even though it is regulatory in its impact. Scientists, like lawyers or doctors, have reason to prefer a self-regulatory process to external controls imposed by the government. A properly functioning review system not only validates knowledge-claims, but also helps the scientific profession to maintain its monopoly over scientific knowledge and over the allocation of funds for the generation of new knowledge. Almost by definition, peer review reaffirms the proposition that only scientists are qualified to judge the validity of work done by their professional peers. Effective self-policing thus enhances the autonomy and social prestige of science,[6] while holding scientists accountable only to standards considered reasonable by their coprofessionals.

Peer Review in Practice

Although peer review is ubiquitous in science,[7] the term corresponds to no single, universally accepted set of practices. Procedures used by journals are substantially different from those of grant-giving agencies, and reviewing traditions within these institutions vary widely. From the standpoint of the present discussion, probably the most significant point of commonality among peer-review systems is that the ultimate decisionmaker, whether government official or journal editor, is seldom bound to follow the reviewers' recommendations exactly.

As Stephen Lock, editor of the *British Medical Journal (BMJ)*, has noted, review procedures used by most scientific journals center around the use of multiple external referees, who are selected—and to varying degrees guided, instructed, and manipulated—by the journal's editor.[8] Common differences in editorial peer-review practices relate to such matters as the number of referees and the nature of the guidance provided to the referee by the editor. Another major area of divergence is the degree of confidentiality in the review process. Many journals favor "double-blind" review, in which neither the author nor the reviewer learns the other's identity. However, it is not uncommon for editors to reveal the author's identity to the reviewer, and signed reviews are the norm for some scientific journals.

Peer review of research grants is a much more recent phenomenon than editorial peer review, though the awarding of unreviewed grants by the federal government today seems even more unthinkable[9] than the publication of an unreviewed article in a leading scientific journal. Like scientific journals, federal grant-making agencies are interested first and foremost in the merits of the proposed work: its originality, methodological soundness, and likely contribution to knowledge. The awarding of research grants, however, is influenced by extrinsic considerations as well. Grants may have to be disbursed in accordance with broad governmental policies favoring particular areas of research. Moreover, as custodians of public funds, granting agencies are accountable to Congress and ultimately to the public for their expenditures on R&D. Their review mechanisms accordingly are designed with an eye to both scientific and political legitimation (the former for technical quality, the latter for overall purpose) and they are thus highly relevant to the needs of regulatory agencies. Federal research-sponsoring organizations, for instance, have adopted rather more formal peer-review mechanisms than most journals, and they generally seek advice from panels of experts rather than from a few select individuals.

At the National Institutes of Health (NIH), peer review is extensively used both to maintain the quality of intramural research and to manage NIH's vast extramural research programs. Review occurs in two stages.[10] The Institutes' central Division of Research Grants (DRG) receives and records applications, which are then referred for review on their scientific merits to one of NIH's specialized study sections. The sections are organized along disciplinary lines and consist entirely of nongovernmental scientists. Most study sections serve more than one of NIH's institutes or research divisions, and thus are not captive to a particular set of funding preferences. Priority scores assigned to each application by the relevant study section are the single most important factor in NIH's funding decisions. However, project applications also undergo a second round of review by the

National Advisory Council attached to each NIH research unit. These bodies include laypeople as well as experts, and their task is to ensure that grants conform to public policy criteria in addition to purely scientific norms. The final funding decision is made independently by each NIH unit.

The NIH review process has served as a model for several smaller agencies, such as the Alcohol, Drug Abuse, and Mental Health Administration, which adopted a reviewing system, following major public controversy, as recently as 1979.[11] EPA too has drawn on the NIH model, although, consistent with its extremely varied R&D initiatives, EPA uses a number of different types of peer review. For example, the agency relies on its Science Advisory Board to review its research programs broadly, while more directed forms of peer review are used in connection with specific intramural and extramural research projects.[12]

The National Science Foundation (NSF) employs a variety of reviewing mechanisms for discriminating among proposals for basic scientific research. While NIH uses standing panels, NSF more commonly relies on ad hoc panels of referees selected by its program managers to represent specific areas of expertise. The basic format for peer review at NSF is to send proposals out to selected external reviewers, who are asked to rate them along several dimensions, including the scientific merit and significance of the proposal and the qualifications of the principal investigator and the supporting institution. Funding decisions are made within each NSF program either directly on the basis of the referees' written evaluations or after a meeting in which an advisory panel further assesses the relative merits of a group of proposals. Details such as the number of reviewers (usually at least five) and the guidance provided to referees vary from one program to another and according to the nature of the proposal, for example, whether it is an unsolicited application or solicited in furtherance of a particular NSF research initiative.

Unlike the jury verdict in a courtroom trial, peer review is not binding. Both journal editors and managers of federal research programs retain discretion to override the recommendations of their reviewers, though the methods they use and the degree of flexibility they enjoy are not precisely comparable. The need to work with fixed, formal reviewing mechanisms to some extent restricts the discretion of funding agencies. But even in agencies like NIH and NSF, program managers can influence the course of peer review through careful choice of referees.

Journal editors on the whole enjoy even greater freedom to adjust the processes of peer review to meet particular editorial goals and policies. At the *New England Journal of Medicine* (*NEJM*), an influential publication addressing a broad cross-section of the medical research community, peer review is carefully managed so as not to rob editors of their decisionmaking authority.[13] As a first step, editors partially predetermine the outcome of peer review through the selection of referees. A senior editor at *NEJM* notes that potential reviewers generally conform to a few familiar types—the main competitor, the severe critic, the statistician—and experienced editors often know in advance what kinds of comments to expect from such individuals. *NEJM*'s referees are specifically requested not to make judgments about the publishability of a paper, since this is considered the prerogative of the editors. Reviewers' comments submitted to *NEJM* may be censored in transmission to the author so as to safeguard editorial discretion still further. For example, superlatives, whether positive or negative, may be deleted, especially when editors decide not to follow a referee's recommendations. Finally, in contrast to research funding agencies, journals like *NEJM* always approach their referees individually, for to do otherwise might produce a peer consensus that editors would find difficult to override.

Why is there such reluctance to let peer review play a more determining role in scientific publishing? The dominant reason offered by editors of major journals is that they are responsible for balancing more interests than are evident to individual reviewers. *BMJ*'s Stephen Lock notes, to begin with, that editors have a different set of responsibilities with respect to the scientific literature than do most referees.[14] The editor's primary duty is to promote the circulation of ideas not found trivial, unoriginal, or wrong. Hence editors may be interested in enhancing or strengthening promising pieces, whereas referees may be more inclined to dwell on their technical defects. Editors also have to respect their journal's overall mission by making sure that published papers address the right audience and represent an appropriate mix of views and interests. Moreover, editors can and should make comparative judgments that referees are not equipped to make. For example, is the article under consideration significant compared with other work in the field? Is there a better or more comprehensive piece on the same subject waiting in the pipeline?

Editorial discretion is also indispensable for tailoring review procedures to the nature of the material under consideration. At *NEJM*, for example, review articles are refereed more extensively than ordinary scientific papers even when they are solicited by the journal. Because these are high-impact pieces, *NEJM* may request comments from three to four referees, instead of the usual two, in order to obtain the best possible scientific consensus. Controversial articles—those which challenge received views, for example, or could have a significant impact on policy—are often more carefully scrutinized than other papers, though this may simply be a consequence of the fact that such work tends to get mixed reviews and hence needs more follow-up. Thus a study of success rates in small-scale surgery for breast cancer awaited publication at *NEJM* for more than a year while the journal had it thoroughly reviewed.[15]

An even clearer instance of editorial peer review taking its cue from policy considerations occurred in connection with a cancer study performed by Carl J. Johnson of the Medical Care and Research Foundation in Denver. Johnson's research purported to find an excess incidence of cancer among Mormons exposed to nuclear fallout from weapons testing in southern Utah. The study was first publicly discussed in a lawsuit against the federal government, when it was introduced in evidence by cancer victims claiming compensation for alleged fallout damage.[16] At the trial, the Johnson study was attacked by experts testifying on behalf of the government, but its credibility received a boost when the prestigious *Journal of the American Medical Association* (*JAMA*) published it in early 1984. George Lundberg, the editor of *JAMA*, announced that he had devised a special peer-review procedure for the Johnson article, consisting of "three cycles of appropriate, anonymous peer review." Explaining this extraordinary procedure, Lundberg said, "I had to be absolutely sure it was responsible and valid because it flies in the face of Government policy and has such a huge potential impact."[17]

As these examples indicate, even in the bastions of scientific research and publication, peer review serves a mixed and multiple function. Although its primary purpose is to provide quality assurance, peer review is also used more or less consciously by both editors and granting agencies to further social objectives, from upholding a funding program's legislative mission to providing support for litigation. An examination of review procedures thus reinforces the observations made in Chapter 1 about the socially constructed character of science's cognitive authority.

Instructive Failures

Just as technological disasters disclose faults in the social systems for producing and managing technology,[18] so the failings of peer review expose the social processes and social conflicts that underlie scientific claims. Skeptics within and outside the scientific community have long acknowledged the gap between actual and ideal in peer review. A much-quoted letter of Thomas Huxley, an early and distinguished critic, remarked on "the intrigues that go on in this blessed world of science" and asserted that if a paper of his were to be referred to a particular colleague, "he will pooh-pooh it to a dead certainty."[19] Similarly, Sir Theodore Fox, a former editor of the British medical journal *Lancet,* was so skeptical of peer review that he wondered if there were any systematic differences between the articles his journal published and those it rejected.[20] Other less prominent scientists admit in private that peer review is not a fail-safe guarantor of quality in science but merely "the best we have." Explanations for these unenthusiastic appraisals can be found in the critical literature on peer review.

Bias and Inconsistency

Huxley's suspicion that referees can be motivated by personal bias is widely shared by scientists, especially by those denied recognition or funding. Others fear that reviewer decisions may be skewed by preferences for particular institutions or geographical areas, prejudices related to gender, scientific intolerance, or conflicts of interest between reviewer and researcher. Evidence of such bias is sketchy, since there have been remarkably few systematic studies of peer review, but it supports the hypothesis that reviewers are neither wholly objective nor wholly consistent in assessing the work of their professional compeers.

A provocative experiment to test the importance of institutional bias was carried out in the 1970s by two psychologists, Douglas Peters and Stephen Ceci.[21] The two researchers targeted thirteen leading psychology journals and selected one article from each written by an author from a high-prestige institution and published within the previous eighteen to thirty-two months. The articles were resubmitted to the same journals with some minor cosmetic alterations to their titles, abstracts, and opening paragraphs. In addition, Peters and Ceci changed the names of the authors, but not the sex, and assigned them fictional institutional affiliations. Upon resubmission, only three articles were recognized as previously published work; of the remaining ten, one was accepted again for publication and eight were rejected on grounds of poor study design or other qualitative reasons. Thus in nine out of thirteen cases, neither the editors nor the reviewers detected the deception or the paper's unoriginality. Peters and Ceci concluded that there were two possible explanations: either the reviewers selected in the second round of review were less competent than those who read the original articles, or there was systematic bias favoring well-known institutions over less prominent ones.

These results, of course, are scarcely conclusive. Other studies of peer review have found no correlation between the author's institution and the acceptance or rejection of articles,[22] and there may be explanations other than bias for the outcome of the Peters and Ceci experiment. For example, since the targeted journals had high rejection rates, the refusal to accept articles on resubmission could have been merely random rather than a result of systematic bias. In any event, because of the small number of test cases and other design flaws, the Peters and Ceci study should most appropriately be viewed as a hypothesis-generating piece that demands further confirmatory research.

Bias against unknown investigators has been documented to some degree, partly by way of anecdotes concerning the rejection or belated collegial acceptance of important papers by obscure scientists.[23] There is more systematic evidence for a phenomenon that Merton termed the "Matthew effect," whereby the accomplishments of scientists who have won high distinction (for example, the Nobel prize) tend to be overestimated, while the contributions of lesser-known colleagues are underestimated.[24] In parallel work, Zuckerman has shown that Nobel laureates tend to give first authorship of joint articles to co-workers, possibly in order to offset the Matthew effect.[25]

Another possible form of bias is the inclination to favor particular research orientations. In one suggestive study, seventy-five referees were requested to review a brief article on issues in behavior modification.[26] The bibliography, introduction, and research methods section were kept identical for all reviewers, but the data and discussion sections were modified either to conform to the referee's own research perspective or to oppose it. In some cases, these sections were left incomplete and the referees were told that they were still in preparation. A statistically significant number of reviewers rated the paper's methods, presentation and discussion more highly when the research orientation was in accord with their own than when it was not. Ratings were even higher when the paper included no results. These findings led Lock to observe that "for all disciplines [there] is the necessity for choosing reviewers with different standpoints, both vis-à-vis the author and one another."[27]

Scientific referees may be inconsistent in their evaluations even in cases where no identifiable bias is at work. This was the conclusion of an extensive study of NSF peer-review practices undertaken by Stephen Cole, Jonathan Cole, and Gary A. Simon.[28] In the first phase of their research 1200 NSF proposals were analyzed, along with reviewer comments on 250 of these. This part of the study uncovered no systematic biases in the review process. In particular, reviewers from major institutions did not appear to give preference to proposals from other major institutions. There was a low to moderate correlation between reviewer ratings and the rank and location of the department and the rank of the investigator in the department.

The second phase of the study, however, sought a reevaluation of 150 proposals covering three fields (solid-state physics, chemical dynamics, and economics). The twelve referees selected for the purpose were given identical instructions and were selected from the same population of referees as in the original round of evaluations. In 24–30 percent of the cases, the second assessment reversed the first. The investigators concluded, therefore, that chance plays a large part in the assessment process: the fate of an application is only half determined by the characteristics of the proposal and the principal investigator, while the other half depends on the "luck of the reviewer draw."[29] This disagreement among assessors was attributed by the Coles and Simon to a basic lack of consensus in both the social and natural sciences as to what "good science" in the field is or should be.

Credulous Review

The idealized account of peer review holds that it is not only dispassionate (that is, free from bias), but also critical, requiring referees to look with a cold eye on a paper's originality, methodological reliability, and deductions.[30] But lack of originality, plagiarism, methodological errors, and faulty inferences have all been known to escape the notice of scientific reviewers, even when these were gross enough to suggest fraud or misconduct.[31] Should we regard these incidents as accidents, or do they reveal deeper forces at work in science that inhibit critical scrutiny by peers?

It is a common belief among scientists that the level and intensity of peer scrutiny are closely related to the interest of the result. Pathbreaking claims, such as the discovery of high-temperature superconductivity or "cold fusion,"[32] generate immediate attempts to dissect and replicate the claimed results, both before and after formal publication. Deception and error, scientists plausibly argue, are virtually impossible to maintain in such an atmosphere. By contrast, deceivers may get away indefinitely with various forms of misconduct when their work is not interesting enough to attract the attention of competitors. This is arguably what explains the success of Elias Alsabti, a medical student of Iraqi origin, who plagiarized up to sixty scientific papers, publishing them again in relatively obscure journals in his own name, sometimes accompanied by fictional coauthors.[33] Similarly, plagiarism in a number of review articles by the Harvard psychiatrist Shervert Frazier escaped detection for up to two decades possibly because they did not purport to present original research data.[34]

Yet methodological problems and fraudulent data sometimes slip through the net of peer review even when the scientist is well known and the work is both important and highly regarded. The case of Cyril Burt, an influential British psychologist, illustrates the point.[35] When Burt's theories about the heritability of intelligence came under attack in the 1950s, he began publishing a series of articles to defend his views. The centerpiece of this work was a set of studies purporting to show that identical twins raised in different environments nevertheless have similar IQs. The work was allegedly based on research carried out twenty to thirty years earlier, but updated with the assistance of two colleagues. The data proved highly influential with U.S. psychologists such as Arthur Jenner and Richard Herrnstein, who also espoused the hereditarian view.[36]

It was not until after Burt's death that Leon Kamin, a Princeton psychologist, became suspicious of the extraordinary consistency of the results, as well as the absence of precise information on Burt's research methods. Kamin eventually concluded that Burt's work on identical twins had been motivated by the desire to prove the hereditarian case and that the data he left behind were "simply not worthy of our current scientific attention."[37] This assessment was seconded a few years by Leslie Hearnshaw, Burt's official biographer, who found evidence of outright deception in his diaries and correspondence.[38] The twin data apparently included some invented cases, and the two coauthors who helped update the material seem never to have existed. How could someone of Burt's eminence have practiced such deceptions and got away with them in his lifetime? In explaining the case, William Broad and Nicholas Wade concluded that "science is not self-policing. Scholars do not always read the scientific literature carefully. Science is not a perfectly objective process."[39]

The story of John Long, a noted researcher at the Massachusetts General Hospital, also suggests that scientific claims may be accepted all too easily when the worker is affiliated with an elite institution. Long built a flourishing career between 1970 and 1979 on research into the biochemistry of Hodgkin's disease.[40] Most of his work was based on an unusual feat: his success in establishing four permanent Hodgkin's cell lines, each allegedly derived from a different human patient. Long's deception came to light only when a colleague, in this case a junior associate, became suspicious of his superior's unaccountable success with a measurement that he himself had not been able to carry out successfully. A lengthy, painful inquiry established that Long's four cell lines were derived not, as he claimed, from Hodgkin's disease patients, but from a Colombian owl monkey and from one human patient who had never had any observable tumors associated with the disease. None of the referees who reviewed Long's successful grant applications and published articles had succeeded in detecting these gross deceptions, although it was common knowledge that permanent, uncontaminated Hodgkin's cell lines are extraordinarily difficult to establish. There is thus considerable irony in the comments offered to Congress by Ronald Lamont-Havers, then research director of the Massachusetts General Hospital, at a hearing on fraud in the biomedical sciences:

> It is very difficult to fudge data so that it will be uncritically accepted by an interested and knowledgeable audience. The greatest protection in science is the critical review and analysis of the published data and procedures by the scientist's peers . . . The key element is critical review of the scientist's peers.[41]

Misrepresentations in the reporting of data are in fact less likely to be detected by peer reviewers than by co-workers in the same laboratory, who often are the first to sense such problems as extraordinary speed or too great neatness in results obtained by their fellow scientists.[42] But even when someone's work appears too good to be true, colleagues may be lulled into believing that the individual has unusual gifts. So, for example, the Cornell University biochemist Volker Vogt was not immediately alerted to the possibility of fraud when frustrated in his efforts to replicate the experiments of Mark Spector, a promising graduate student in his department.[43] Spector's standing collapsed only when Vogt accidentally obtained some of the original experimental preparations and realized that they had been "cooked."

Science by Convention

The most interesting insights to be drawn from the peer-review literature, however, relate not to the aberrant practices of individual scientists but to the day-to-day workings of science. In this connection, the career of John Darsee, a brilliantly promising researcher at Harvard Medical School, proved especially revealing.[44] Darsee was a protégé of Eugene Braunwald, a renowned and influential cardiologist, physician-in-chief at two Harvard-affiliated hospitals, and director of two separate research laboratories. Darsee's own productivity surpassed even the expectations of his high-powered laboratory chief; in just two years he published some one hundred papers and abstracts, many coauthored with Braunwald. This phenomenal activity, however, aroused the suspicions of Darsee's co-workers, who could not understand how the young researcher had the time to carry out so many experiments. In 1981 shocked colleagues discovered Darsee faking data for a study that was about to be published. He admitted the fraud, but denied ever having engaged in such practices before. As a result, he was permitted to stay on at Harvard and continued working there for several months until new problems were uncovered in experimental data he had reported to NIH under a federal research grant. Subsequent investigations revealed an extensive record of misconduct, begun even before Darsee joined Braunwald's laboratory, which amply confirmed his co-workers' suspicions that the observed instance of faking was not an isolated one.

As Darsee's principal supervisor, Braunwald was heavily criticized for fostering an atmosphere of competitiveness coupled with carelessness that invited misconduct to flourish. Much more seriously, however, Braunwald was himself charged with misrepresenting the results of his joint research with Darsee. The accusation was brought by two

NIH scientists, Walter W. Stewart and Ned Feder, whose concern about scientific misconduct and faulty peer review had led them to reanalyze Darsee's published papers.[45] Based on their unusual survey, Stewart and Feder concluded that Darsee's publications contained errors and inaccuracies that should have been detected by his coauthors and peer reviewers.

One of the most damaging allegations made by the NIH scientists was that Braunwald and Darsee had used some historical controls in one set of animal experiments and had knowingly failed to report this fact in print. According to the published articles, all the dogs involved in a study of treatments for heart damage had been randomly assigned by computer to a treatment group or a control group. In reality, the control group included a mix of contemporary controls and randomly selected controls from previous experiments using dogs under similar, though not identical, conditions. Interviewed by the *New York Times,* Braunwald admitted the practice, but justified the use of mixed controls as a means of avoiding unnecessary sacrifices of test animals.[46] He also denied that reporting the method of selecting the controls was vital for scientists wishing to replicate his experiments.

Approached by the same interviewer, however, Benjamin Lewin, editor of the prestigious journal *Cell,* flatly rejected Braunwald's position, asserting that it is "absolutely critical" to describe controls accurately. Lewin saw the accurate reporting of controls as indispensable not only for the sake of a study's replicability but also to ensure that it is given due weight by other researchers.[47] Since experiments using mixed controls are considered less powerful than those with only contemporary controls, failure to indicate which approach was used could be quite misleading.

This standoff about acceptable research practices was not officially "resolved" until the Stewart and Feder article was published in the British journal *Nature* in January 1987.[48] In an accompanying rejoinder, Braunwald again defended the use of mixed controls. His laboratory, he now explained, had a standing policy of randomizing control animals. In a series of similar drug-intervention experiments by the same researcher, contemporaneous controls would be used only for the first intervention. Thereafter, the researcher would use a mixture of historical and contemporaneous controls in order to avoid "unnecessary sacrifice of animals."[49] The editor of *Nature* accepted this explanation, noting with only mild disapproval that "while

Braunwald's defence of the use of 'historical controls' is entirely proper, it remains a fact that the procedure that should have been followed . . . is not adequately explained."[50]

The exchange among Braunwald, his detractors, and his defenders provides fascinating insights into the socially constructed nature of experimental validity. The episode illustrated, in the first place, that respectable scientists with different disciplinary backgrounds (Braunwald, a cardiologist; Lewin, a molecular biologist) can disagree profoundly on something so basic to "good science" as the appropriate method of selecting and reporting on controls.The possibility of such disagreement was apparently not recognized by scientists reviewing Dante Picciano's study at Love Canal. Second, the controversy underscored the conventional character of methodological choices made by experimenters. Braunwald's laboratory policy of randomized controls appeared to be dictated in this case by social policy considerations—the need to save animals—rather than by any intrinsic property of the experiment. Finally, *Nature*'s decision to publish Braunwald's explanation and to accept it as "entirely proper" became a boundary-drawing exercise designed at once to placate the cardiologist's critics and to restore his legitimacy as a scientist. Readmitted to the pages of *Nature,* Braunwald in effect was readmitted to science, while his former student Darsee was designated as genuinely deviant and consigned to the outer darkness.

Regulatory Science: Content and Context

The argument for importing peer review into the regulatory process rests, as we have seen, on the secondary premise that regulatory science (science used in policymaking) is no different from science in the research setting. On its face, however, this premise appears to conflict with assertions by science policy analysts that regulatory (or mandated) science does have properties that distinguish it from research science.[51] The attempt to draw clean distinctions between various subtypes of science (such as, pure vs. applied) is known to be fraught with conceptual difficulties.[52] Yet most users and observers of science concur that there are some scientific "activities and roles ordered around the extension of knowledge and competence without any regard for practical application. There are other roles and activities concerned solely to increase and improve the stock of existing practically useful techniques, processes and artifacts."[53] Regulatory science, however defined, falls on the practical side of this division, for its purpose clearly is to produce "techniques, processes and artifacts" that further the task of policy development. The connections between this property of regulatory science and the technocratic model of legitimation need to be more fully explored.

Science used in the policy process differs from research science most significantly in its context, but also partly in its content. With regard to content, regulatory science can best be thought of as the aggregate of three different types of scientific activity. In the first place, regulatory science includes a component of *knowledge production*. Studies designed to fill gaps in the knowledge base relevant to regulation may either be performed or sponsored by regulatory agencies or they may be carried out under ordinary conditions of academic research. In either case, as we saw in Chapter 2, their status as "science" is frequently open to question. Second, regulatory science includes a substantial component of *knowledge synthesis*. As Salter notes, secondary activities, such as evaluation, screening, and meta-analysis, play a much larger role in regulatory science than open-ended, original research.[54] The products of regulatory and research science accordingly are rather different. Whereas research science places greatest value on published papers, certified by peers as true, original, and significant, science conducted for policy is rarely innovative and may never be submitted to the discipline of peer review and publication.

The third and most contentious branch of regulatory science is *prediction*, the activity that requires the decisionmaker to determine how serious or significant a risk is created by a regulated technology. Because prediction involves so many elements of uncertainty and discretionary judgment, analysts seeking to understand regulatory science have paid it disproportionate attention. For example, Alvin Weinberg, a physicist and commentator on science-based regulation, suggested that the predictive component of decisionmaking should alone be treated as "a new branch of science, called regulatory science, in which the norms of proof are less demanding than are the norms in ordinary science."[55] The proposal that the scientific basis of regulatory decisionmaking should be validated in accordance with "less demanding" standards from those applicable to "ordinary science" is strikingly reminiscent of the observations made by the D.C. Circuit Court in developing the science policy paradigm. As we saw in the preceding chapter, however, this approach to thinking about regulatory science proved largely ineffectual for purposes of reducing policy conflict. It merely opened the way for politically loaded boundary work; actors dissatisfied with policy outcomes gained easy political leverage by labeling regulatory science as "bad science."

From the standpoint of political legitimation, differences in the methodologies and end products of research science and regulatory science are not nearly so significant as the differences in their institutional and cultural environments. In the end, it is these contextual differences that shape and limit public perceptions about the validity of regulatory science.

One of the most telling features of regulatory science is the relatively heavy involvement of government and industry in the process of producing and certifying knowledge.[56] Science carried out in nonacademic settings may be subordinated to institutional pressures that critically influence researchers' attitudes to issues of proof and evidence. These orientations in turn affect the packaging and presentation of scientific results. As will be seen in later chapters, opinions about such issues as the need for contemporaneous controls or the lack of statistical significance seem to depend systematically on the observer's institutional affiliation.

Regulatory agencies, moreover, are accountable to numerous nonscientific watchdogs: Congress, the media, the courts, and the interested public. Political demands imposed by these external supervisors can have an impact even on the consideration of technical issues. Regulators may be required, for instance, to speed up their timetables for gathering and assessing evidence in response to public concerns about risk. Time thus becomes an important differentiating factor in the environments for regulatory and research science. While scientists working in a "pure" research setting have relatively unlimited time (subject to funding and career constraints) for testing hypotheses or proving conjectures, scientists working to meet policy needs are under constant pressure to deliver results quickly. In the regulatory context, a decision to wait for more data amounts to (or is perceived as) a decision not to act. Hence scientists involved in policymaking frequently find that they cannot credibly avoid accepting or rejecting a conclusion on the ground that the matter calls for further study.

A further important difference between the two areas of scientific activity is in the definition of the standards by which each is evaluated. Academic researchers, on the whole, work within established scientific paradigms, subject to relatively well-negotiated prior understandings about what constitutes good research methodology. Borrowing an image from French sociologist Bruno Latour, we can say that scientists in academia make more extensive use of "black boxes"—facts or claims that other scientists view as too impregnable to be worth contesting or deconstructing.[57] Regulatory science, by contrast, is more often done at the margins of existing knowledge, where science and policy are difficult to distinguish and claims are backed by few, if any, allies or black boxes. As a result, fully specified, seemingly impersonal criteria for judging the findings of regulatory science are seldom available in advance. Instead, the guidelines for validating science in the regulatory context tend to be fluid, controversial, and arguably more politically motivated than those applicable to university-based research.

The salient differences between research science and regulatory science are summarized in Table 4.1. They help explain why the standards by which agency experts judge the acceptability of science may diverge from those used by research scientists. Such discrepancies, in turn, open the door to disputes among regulators, industry, and the scientific community about the validity of work done for purposes of supporting policy.

Implications for Regulatory Peer Review

The foregoing discussion calls into doubt some of the basic premises of the technocratic reformers of regulation. Peer review appears not to be the objective, dispassionate process that its advocates represent it to be. Standards of validity in science are also revealed as somewhat fragile constructs that may hold up under friendly scrutiny, but are apt to disintegrate under controversy or critical review. Accordingly, the notion that science-based regulation can be lifted above politics and ideology through peer review appears seriously misguided. Many of the factors that diminish the credibility of peer review in the research setting seem likely to exert an even more negative influence in the regulatory environment.

The potential for bias, to begin with, is more pronounced in the context of regulatory science than in research science. Scientists serving on peer-review panels are frequently selected from elite scientific institutions; they may therefore approach their task with an inherent bias against work produced by agency experts, who are believed by many to be less competent on average than scientists actively pursuing a research career. Communication between scientific reviewers and

Table 4.1. Regulatory science and research science.

	Regulatory science	Research science
Goals	"Truths" relevant to policy	"Truths" of originality and significance
Institutions	Government Industry	Universities
Products	Studies and data analyses, often unpublished	Published papers
Incentives	Compliance with legal requirements	Professional recognition and advancement
Time-frame	Statutory timetables Political pressure	Open-ended
Options	Acceptance of evidence Rejection of evidence	Acceptance of evidence Rejection of evidence Waiting for more data
Accountability		
Institutions	Congress Courts Media	Professional peers
Procedures	Audits and site visits Regulatory peer review Judicial review Legislative oversight	Peer review, formal and informal
Standards	Absence of fraud or misrepresentation Conformity to approved protocols and agency guidelines Legal tests of sufficiency (e.g., substantial evidence, preponderance of the evidence)	Absence of fraud or misrepresentation Conformity to methods accepted by peer scientists Statistical significance

regulatory scientists may also be impeded by a clash of cultures between regulatory and research science.[58] These effects may be partially offset by the agency practice of commissioning scientific studies from highly credentialed researchers at academic institutions or from private consulting groups with specialized expertise and established reputations. In such cases, institutional bias should not be an important distorting factor, although the competence of experts selected by an agency can always become an issue, as in the disputes over Paul

Newberne's nitrite study or Picciano's study of chromosome damage at Love Canal.[59]

Bias related to research perspectives can be expected to play a more serious part in the review of regulatory science by outside experts. Many of the experimental methods and analytical techniques used in assessing health and environmental risks remain controversial and are not based on a broad scientific consensus. Thus, a particular peer reviewer's disciplinary training and professional affiliations could well have a bearing on his or her evaluation of risk assessments carried out by or for the regulatory agencies. Such effects could be magnified through a conscious decision by the sponsoring agency to exclude or overrepresent a particular department or research tradition on an expert review panel.[60]

Studies of risk assessment and risk management in different countries provide empirical evidence that research perspectives can influence the way scientists approach the evaluation of hazards. The risk of cancer from exposure to environmental chemicals, for example, has frequently aroused greater concern in the United States, where the techniques of probabilistic risk assessment are most widely used, than in countries like Britain, where health risk assessment is dominated by epidemiologists rather than toxicologists or biostatisticians.[61] Such biases would presumably carry over into scientific advice and peer review, since scientists skeptical of particular methodological approaches would be disinclined to accept their use by regulators.

The most obvious source of bias in the regulatory environment, however, arises from the fact that expert referees may either be formally affiliated with particular interest groups or otherwise have a stake in the outcome of the regulatory process. It has been amply documented that technically trained adversaries can exploit uncertainties in the scientific knowledge base to construct evaluations consistent with their political objectives.[62] Since the early 1970s, for example, environmental and labor groups have been at loggerheads with industry experts over how best to interpret animal tests using high exposure doses, how to evaluate benign tumors in relation to malignant ones, how to assess tumors for which there is a high background rate in the tested species, and so forth. In theory, agencies could stack the deck in favor of one or another viewpoint by simply selecting peer reviewers with known opinions on these issues.

The wish to avoid such bias in regulatory peer review (and, by extension, in scientific advice) is universally expressed by both agency officials and scientists. Legal safeguards such as open decisionmaking and judicial review provide additional barriers against the manipulation of expert opinion simply to further specific political interests. As will be seen in the next chapter, however, attempts to negotiate scientific differences openly, by making expert advisory committees more politically balanced, have encountered substantial opposition.[63]

The problem of inconsistency in peer review likewise may affect regulatory science more adversely than research science. Inconsistent reviews of unpublished papers do not necessarily impede the discovery and communication of scientific knowledge. Stephen Lock's study of editorial peer review, for example, indicated that refereeing prevents publication only in a small percentage of cases; more often, peer review simply channels scientific papers to different, possibly more appropriate, outlets.[64] With respect to regulatory science, however, peer review is likely to perform a more obviously gatekeeping function. Agencies subject to supervision by scientific experts will seldom feel confident enough to act unless their scientific assessments survive peer review. Accordingly, avoidance of inconsistency has to be a major concern in regulatory peer review. Widespread perception that outcomes are determined simply by "the luck of the reviewer draw" could fatally damage the credibility and legitimacy of regulatory policy.

In research science, time is usually on the scientist's side. Conclusions do not have to be accepted as true until most members of the relevant community are satisfied by the available evidence. And if mistaken or fraudulent results gain temporary acceptance, their chances of escaping detection over time are low, especially for research making significant scientific claims or announcing breakthroughs. Time, however, is not an ally of regulatory agencies, who ordinarily must make decisions before a consensus forms about the acceptability of evidence. Temporal constraints also magnify the likelihood of error, thus increasing the agencies' vulnerability to charges of incompetence.

Regulators enjoy some advantages over scientific referees in their efforts at preventing outright misconduct, but there are associated costs as well as benefits. On the plus side, agencies have real, though limited, resources for inspecting and monitoring laboratories where research relevant to policy is in progress. As Broad and Wade suggest in their study of fraud in science, such on-site supervision may be the most powerful instrument available for uncovering scientific malpractice.[65] Regulatory science also tends to be published and debated under more adversarial and less credulous circumstances than research science. Manipulated results or statistical errors are thus less likely to pass undetected through the decision-making process. On the minus side, however, a contentious regulatory environment may promote indefinite deconstruction of scientific claims and may limit the possibilities for negotiated settlement of technical arguments.

Finally, the technocratic reform movement overlooks the fact that peer review is not the only mechanism by which scientists build up their cognitive authority. Informal supervision through professional networks in and out of the laboratory, and through scientific "grapevines" tended by journal editors and other gatekeepers, supplement peer review and help maintain consensus and credibility in science. Agencies, for their part, are governed by a network of administrative laws and procedures that hold decisionmakers to high standards of rationality, impartiality, and openness, thereby conferring legitimacy on their use of science. At the same time, agencies are subject to constraints that cut in the opposite direction. Mandates to act precautionarily or within specified time limits, for instance, are often antithetical to agency efforts to project an image of technical impartiality and expertise. The next six chapters document a variety of ways in which agencies and their scientific advisers have balanced these countervailing pressures in the course of formulating science policy.

5

EPA and the Science Advisory Board

The changing role and responsibilities of the Science Advisory Board (SAB) since its creation in 1974 mirror EPA's maturation as user and generator of regulatory science. Along with the agency, the Board was drawn into some of the most controversial arenas of science policy in the mid-1970s, but its intervention in the regulatory process at this time was unsystematic and lacking in definition. By the end of the 1980s, however, external indicators all suggested that the relationship between EPA and the SAB had settled into a pattern of successful cooperation. During the second Reagan term, in particular, the Board's resources and reputation grew, along with its breadth of involvement in EPA's decisionmaking. This evolution provides an excellent case study for exploring the central conundrums of scientific advice: how has an advisory committee attached to a chronically embattled agency maintained its independence and scientific credibility? How, in particular, has it contributed to science policy without falling captive to EPA, the White House, or organized political interests?

Early Political Challenges

The Science Advisory Board was established by the EPA Administrator in January 1974 to review the research programs of the agency's Office of Research and Development (ORD). In 1976, the SAB was removed to the Office of the Administrator, where it currently resides. Two years later, Congress reestablished SAB on a statutory footing through an amendment to the Environmental Research, Development, and Demonstration Authorization Act.[1] Separate statutory authorization was provided under the Clean Air Act for the Clean Air Scientific Advisory Committee (CASAC),[2] which reviews the scientific basis for National Ambient Air Quality Standards (NAAQS) and is administratively organized as one of the SAB's permanent standing committees. Table 5.1 summarizes SAB's responsibilities as defined by its charter, which is reissued every two years in accordance with FACA requirements.[3]

It did not take EPA many years to discover that the SAB was a dangerous ally, capable of functioning either as a scientific and political troubleshooter or as a lightning rod for controversy. The possible risks of seeking advice from the SAB were brought home quite early in connection with EPA's management of a research program known as the Community Health and Environmental Surveillance System (CHESS), which the agency had inherited from its predecessor, the National Air Pollution Control Administration.[4] Begun in 1967, the CHESS program undertook aerometric measurements of various criteria pollutants, as well as epidemiological studies of the adverse

Table 5.1. Responsibilities of the Science Advisory Board.

Reviewing and advising on the adequacy and scientific basis of any proposed criteria document, standard, limitation, or regulation under the Clean Air Act, the Federal Water Pollution Control Act, the Resource Conservation and Recovery Act of 1976, the Noise Control Act, the Toxic Substances Control Act, the Safe Drinking Water Act, the Comprehensive Environmental Response, Compensation, and Liability Act, or any other authority of the Administrator.

Reviewing and advising on the scientific and technical adequacy of Agency programs, guidelines, methodologies, protocols, and tests.

Recommending, as appropriate, new or revised scientific criteria or standards for protection of human health and the environment.

Through the Clean Air Scientific Advisory Committee, providing the scientific review and advice required under the Clean Air Act, as amended.

Reviewing and advising on new information needs and the quality of the Agency plans and programs for research, and the five-year plans for environmental research, development, and demonstration.

Advising on the relative importance of various natural and anthropogenic pollution sources.

As appropriate, consulting and coordinating with the Scientific Advisory Panel established by the Administrator pursuant to section 21(b) of the Federal Insecticide, Fungicide, and Rodenticide Act, as amended.

Consulting and coordinating with other Agency advisory groups as requested by the Administrator.

health effects of these substances at ambient levels. The program's goal was to create a data base and quantitative risk estimates to support standard-setting on these pollutants.

Concerns about technical problems in the CHESS program and about EPA's interpretation of the data surfaced in 1974, soon after the agency published a monograph analyzing sulfur oxides studies conducted in 1970–71. Questions about the monograph led EPA to consult with the newly formed Science Advisory Board, and in 1975 the SAB issued a report critical of the agency's epidemiological methods and the techniques of presentation used in the monograph.[5] A series of exposé articles published in the *Los Angeles Times* in 1976 added to the agency's troubles by charging EPA staff with deliberate distortions of data to support their regulatory position.[6] Two House committees held joint hearings to investigate these allegations. EPA was exonerated of the fraud charges, but follow-up investigations by subcommittees of the House Committee on Science and Technology concluded that the CHESS program was marred by severe temporal, budgetary, and administrative constraints and that technical errors in the program rendered the monograph on sulfur oxides useless as a basis for regulatory action. CHESS, the investigators concluded, marked at best a "confirmation of previous advances in knowledge" of pollution-induced health effects, and its "lessons could have been learned at much less cost in funds, elapsed time, EPA credibility, and staff morale."[7]

Prefiguring the trend in future science policy controversies, the congressional investigators blamed the monograph's problems on inadequate peer review and recommended the establishment of a panel to monitor all ongoing research projects. They also urged EPA to seek timely peer review and publication of its research results in refereed journals.[8] While endorsing the SAB's involvement in general terms, the report found fault with the Board for providing "only formal criticism, expressed in public meetings attended by press and industry."[9] Relations between EPA and its science advisers, the committee felt, should be restructured so as to ensure less formal, less adversarial, and more forward-looking forms of interaction.[10]

By 1977, a comprehensive survey of EPA's decisionmaking procedures by the National Academy of Sciences implicitly recognized that SAB was needed not only for scientific but for political legitimation.[11] EPA's scientific vulnerability was already apparent. The agency's frequent, though unavoidable, reliance on previously unpublished and unreviewed data exposed it to charges of using bad science. Under the circumstances, some institutional mechanism was necessary to reassure critics of the accuracy and reliability of EPA's scientific determinations. The Academy concluded that the SAB could perform this function by detecting technical biases, errors, or misinterpretations before scientific findings were irretrievably enmeshed in policy deliberations. To enhance the Board's own scientific authority, the NAS study proposed that the office of the SAB chair be merged with that of the administrator's science adviser and that the SAB chair be given more flexibility to initiate review of issues not raised by EPA. The study also recommended keeping the SAB chair's term short, preferably not more than two years, so as to preserve the independence of the office and to make it more attractive to prominent and active scientists.

EPA's political vulnerability, as the Academy study recognized, was in part a structural consequence of building a scientific record through confrontational proceedings. The Academy looked to the SAB as the actor most likely to restore the image as well as the substance of neutrality in such controversies:

> Much of the process by which EPA makes regulatory decisions is adversarial, and often scientific information is provided by one of the principals. Similarly, the Agency itself is sometimes placed in an advocacy role. In either case, review can help to assure a balanced treatment of scientific and technical information.[12]

This analysis contrasted sharply with contemporaneous judicial endorsements of cross-examination and other adversarial procedures as the favored means for testing regulatory science. In retrospect, the Academy's skepticism about such procedures seems a prescient foreshadowing of the decline of court-centered approaches to developing the scientific basis for policy.

But if the National Academy conceived of the SAB as an impartial dispenser of technical rationality, it soon became clear that this view was not shared by others concerned with science policy at EPA. Controversies over SAB's membership during President Reagan's first term showed that major players in the regulatory process were keenly aware of the SAB as a powerful political resource and were quite prepared to see it assume a more activist role in policy. Attempts to control appointments to the SAB became one of the mini-arenas on which pro- and antiregulatory interests played out their deep-seated philosophical and social conflicts.

Environmental interests, by and large, were eager for the SAB to incorporate both lay and scientific views and values, and in 1982 they launched a successful initiative in Congress to diversify SAB's membership. A rider attached to an appropriations bill authorizing R&D funds for EPA demanded that the Board include representation from "states, industry, labor, academia, consumers, and the general public."[13] To justify such explicit interest-balancing, the bill referred to FACA's provision that committees should be "fairly balanced in terms of points of view represented." Congress thus gave FACA's fair balance clause a more liberal reading than most academic and judicial commentators were inclined to do at that time.

But this attempt to substitute a democratic for a technocratic conception of science advice incurred predictable opposition from a White House eager to limit populist influences on environmental policy. President Reagan rejected the appropriations bill just before the 1982 election with a strongly worded veto message reaffirming the political neutrality of science: "The purpose of the Science Advisory Board is to apply the universally accepted premises of scientific peer review to the research conclusions that will form the basis for EPA regulations, a function that must remain above interest group politics."[14] Any other course, the President suggested, would come close to Lysenkoism, a dreaded metaphor for the subjugation of knowledge to political manipulation.[15] This view won enthusiatic support from the highest echelons of the nation's scientific establishment. Both Frank Press, the president of the National Academy of Sciences, and William Carey, the executive director of the American Academy for the Advancement of Science, repudiated the policy of moving EPA's principal scientific advisory committee toward the model of overt interest representation.

An administration committed to deregulation, however, had little interest in converting its rhetorical support for the SAB into dollars for science at EPA. According to a study by the General Accounting Office (GAO) EPA's scientific and engineering staff declined by 6.2 percent between 1973 and 1984, largely as a result of agency-wide controls on hiring imposed after the 1980 presidential election.[16] During this time, the agency's nonscientific workforce increased by 9.4 percent, contributing to a 1.1 percent increase in total personnel. EPA was the only one of seven agencies in the GAO survey where the direction of change in the scientific workforce did not parallel the direction of change in the total workforce. Morale among agency staff dropped as charges proliferated that EPA was suffering a brain drain and was neglecting essential research programs. Anne Gorsuch, the EPA administrator from 1981 to 1983, tried to counter the impression that resource constraints were affecting the caliber of research. "Our emphasis is on quality," she asserted, "not on the amount of money we can spend."[17]

A more direct attack on the SAB came to light in 1983 when the House Committee on Science and Technology made public an alleged "hit list" of EPA's scientific advisers compiled by unknown sources in the initial days of the Reagan presidency. The document listed some ninety scientists on various EPA boards with terse comments on their political ideology and technical competence. The annotations ranged from a curt "get him (her) out" to pointed descriptions of the individual's credentials on environmental issues (for example, "a Nader on toxics"[18] and a "snail-darter type"), which were strikingly at odds with the president's public posture regarding SAB's neutrality. It was unclear whether the list was ever intended as a serious political weapon. Publicly, the agency's own staff, as well as scientists outside the agency, denied that it was ever used in making advisory committee appointments.[19] Scientists given negative ratings on the "hit list," however, were not generally invited back to advise the agency. Further, some construed the consolidation and disbanding of several SAB subcommittees that occurred during this period as part of a Gorsuch plan to purge the Board, although this initiative had actually been begun under Douglas Costle, the previous EPA administrator.[20] In any event, disclosure of the list proved a major embarrassment and provided the impetus for enlarging the SAB to pre-1980 levels, increasing its workload, and expanding the pool of candidates considered by EPA.

A New Cooperation

Following the retrenchment of the early 1980s, SAB's history has generally been one of growth, as Table 5.2 indicates.[21] The Board's resurgence as a significant actor in EPA decisionmaking in the mid-1980s was reflected in an expansion of its membership and in the range and frequency of its scientific reports and reviews. In fiscal year 1986 the number of reviews rose to 65, an increase of 30 percent over the preceding year.[22] They included, besides reviews of individual research programs, an evaluation of the proposed budget for ORD, and targeted reviews of the technical basis for standards, agency policy statements, scientific methodologies, EPA advisory documents,

Table 5.2. Annual data on EPA's Science Advisory Board.

Year	Members	Costs	Staff	Number of reports
1982	31	$ 498,318	9.00	5
1983	37	$ 581,243	7.85	8
1984	41	$ 892,021	12.90	15
1985	54	$1,050,270	12.00	38
1986	53	$1,027,000	12.10	24
1987	67	$1,186,966	12.10	29
1988	67	$1,161,500	14.00	43

and scientific proposals, studies, or surveys. SAB's activities extended across all of EPA's regulatory and research programs; in addition, reviews were conducted for the first time for the enforcement office and one of the regional offices. In its first annual report, the Board assessed these facts and figures as pointing "to a greater understanding of the respective roles and responsibilities by EPA staff, and scientists and engineers that serve on the Board."[23]

The channels by which SAB engages in the review of scientific or technical issues have also diversified. The core of the Board's activities still concerns those regulatory decisions and research programs for which Congress has specifically mandated SAB review. Other issues, however, may routinely be brought before the Board through voluntary referrals by the administrator and deputy administrator or by assistant administrators and their program managers. Requests from the individual program offices are first discussed with the administrator and then submitted to the executive committee for its approval. This internal screening process ensures that only those issues that are genuinely important or difficult are passed on to the SAB for review. Finally, as recommended in the NAS study, SAB now has authority to initiate action on matters where it believes its participation could improve decisionmaking. For example, the Board on its own initiative established a new subcommittee on ecology to help the agency counterbalance what many viewed as an excessive tilt toward human health issues. Structurally, the SAB has evolved over the years into a complex organization operating through a network of committees, subcommittees, panels, and individual consultants. At its head is an executive committee whose chair reports directly to the administrator. As of January 1985, the Board had five permanent standing committees addressing the following issues: environmental health, environmental effects, transport and fate, radiation, environmental engineering, and air pollution (CASAC). Figure 5.1 provides a detailed organizational chart of SAB as of fiscal year 1987, showing its committees and subcommittees, both standing and ad hoc.[24] At any given time, some 200–250 scientists are formally affiliated with the SAB as members, panelists, or consultants. Another several hundred scientists across the country have previously served the Board in one or another capacity.

In the aftermath of the "hit list" crisis, the process of selecting these experts slipped back into the domain of bureaucratic self-management and ceased to be a matter of visible political controversy. EPA regained more or less unquestioned authority to make appointments to SAB and its committees, although the procedure evolved by the agency was far removed indeed from the elite technocratic model recommended by the 1977 NAS study.[25] In place of short rotations of top-flight research scientists, EPA began using repeat assignments and informal interest balancing to acculturate scientists into the special subculture of regulatory science. The search for technically qualified experts was tempered by equally important concerns for breadth, balance, and familiarity with EPA's overarching regulatory mission.

Although EPA has taken pains to open up the SAB appointment process,[26] it remains considerably more discretionary than is the case for many other federal advisory committees, including EPA's own Scientific Advisory Panel for pesticides. Openings are announced in the *Federal Register,* accompanied by general information about the areas of expertise that the agency wishes to have represented. This solicitation is supplemented by letters to professional scientific societies and to individual scientists requesting additional nominations. SAB members actively assist the selection process by proposing potential candidates. Terry Yosie, the SAB's executive director from 1981 to 1988, viewed this multipronged recruitment effort as an effective means not only of enlarging the pool of scientists considered by the agency but of deflecting charges that EPA was building up a narrow body of experts captive to particular scientific or policy orientations.

Conventional conflicts of interest arising from financial ties between members and regulated interests are less of a problem for the Board than for many other advisory committees. With the exception of CASAC, most SAB subcommittees do not advise the agency on regula-

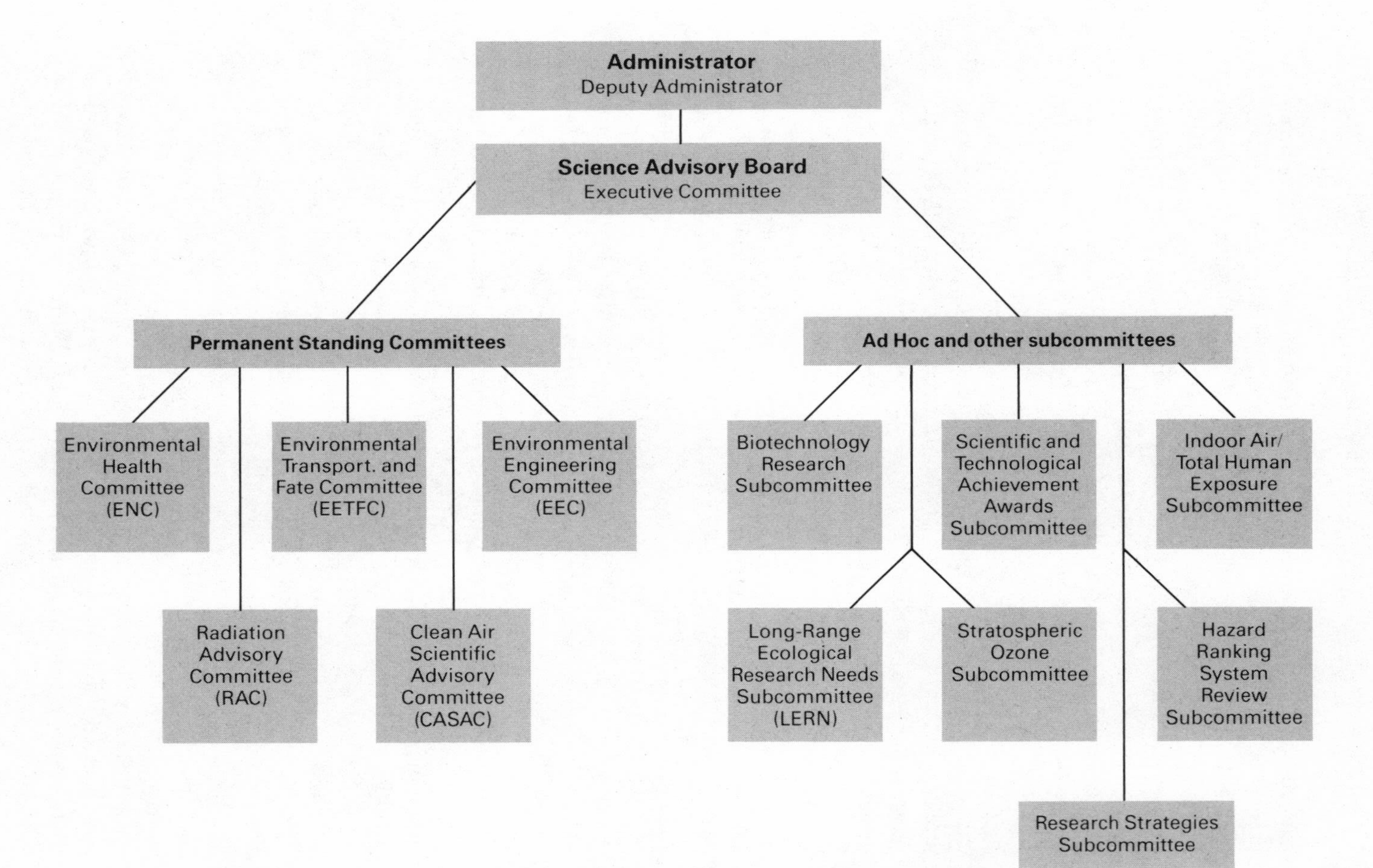

Figure 5.1. Organization of the Science Advisory Board.

tory decisions targeted at particular industries and thus are not in a position to tilt their advice to specific commercial interests. Even for CASAC members, the relationship between advice and policy is quite attenuated, since the NAAQS process affects many different industrial sectors. Nonetheless, the Board has adopted explicit conflict-of-interest guidelines to guard against potentially embarrassing disclosures.[27] Bias arising from a member's disciplinary or research orientation is a real though subtle problem. Advisers selected for their technical expertise are of necessity committed to particular theoretical and methodological approaches, and their receptivity to other schools of thought is likely to be correspondingly limited. While EPA has no formal guidelines to prevent such bias on the SAB, it attempts to steer between either rigidly disqualifying or giving undue weight to experts with clearly defined scientific viewpoints. One method the Board employs is to ask experts to excuse themselves when their own work is being evaluated by other members of the group.[28]

Informally and without fanfare, EPA has used its discretionary powers to transform the SAB into a collaborative rather than competitive source of scientific advice. SAB's membership is largely drawn from academic institutions, but industry, public interest groups, and scientific consulting organizations are also represented. Like the editors of science journals, EPA officials privately admit that an expert's position with respect to issues in regulatory science can generally be predicted in advance. The "neutral" expert is at best a convenient fiction, and balancing varied points of view becomes essential for an agency wishing to foster the appearance of impartiality in its advisory committees. Yosie's policies for staffing the SAB reflected these understandings. While agreeing that formal interest representation would be politically unworkable,[29] Yosie admitted to following a more restrained approach to balancing, using administrative discretion and informal networking to achieve a diversity of backgrounds and interests among experts attached to the Board.[30]

Other aspects of the SAB appointment process (including committee appointments) also indicate that advisers are selected with an eye to much more than merely technical qualifications. Thus, selection criteria are tailored to fit the function an expert is expected to perform. For example, broad scientific experience and an ability to integrate knowledge across disciplines are seen as indispensable assets for Board members, who must function to some extent as science statesmen. Consultants, by contrast, are more often selected for their specialized expertise on narrowly technical issues. It is not uncommon for scientists to be socialized into the advisory culture by serving the Board in several different capacities. For instance, an expert who initially advises the SAB as a consultant may later be recruited as a full-fledged Board member. In this way, it has been possible for EPA to create a network of scientists in sympathy with the agency's peculiar combination of scientific and political needs.

This is not to suggest, however, that EPA crassly manipulates science in order to further its regulatory objectives. Scientists who have served the SAB in one or another capacity acknowledge that what they deliver to the agency is not ordinary science and that extrascientific factors can appropriately be considered in screening experts for service on the Board or its subcommittees. Much of the scientific review required by EPA, for instance, is interdisciplinary in character. Over-specialization is therefore a disadvantage for SAB scientists, who must be able to interact with experts in fields that are quite peripheral to their own. In discussing something as complex as a criteria document, at best three or four committee members may be technically qualified to evaluate specific segments: the toxicological data, the epidemiological evidence, or the biochemistry of the pollutant. As these specialists engage the agency staff in critical discussion, other committee members function almost as a science court, evaluating the overall plausibility of the agency's argument, but not actively participating in its development or refinement.

Further, EPA advisers recognize that the scientific issues raised in environmental decisionmaking seldom touch on the most esoteric or theoretically sophisticated areas within a discipline. The important questions confronted by advisory panels relate as often as not to rather mundane methodological problems that are of little interest to the specialist, such as the adequacy of a study design or the validity of a particular approach to risk assessment. Experienced scientific advisers recognize, as well, that questions to be decided in regulatory science generally involve subjective judgment, and even policy, to compensate for the absence of hard knowledge. What is the "feel" of the evidence taken as a whole? Have the uncertainties been fully identified? Are the studies so inadequate as to be useless, or do they suggest a cause for concern even though they are not perfect? Scientists with narrowly developed areas of expertise are more likely to find such questions bewildering than those who are familiar with a broad cross-section of issues in regulatory science.

From EPA's point of view, then, the best scientific advisers are clearly those with the deepest understanding of regulatory science, especially in the context of EPA's own extremely complex research and policy agenda. Such familiarity can rarely be acquired except through long and close association with the agency, a state that would be almost impossible to achieve under a policy of frequent rotations.[31] Of course, such close ties between the agency and its scientific advisers could foster the impression that the advice being offered is not genuinely "independent." But this is a risk EPA has been willing to take. In particular, the choice of Norton Nelson of New York University to head the SAB during the post-Gorsuch years reflected a recognition on the part of two successive EPA administrators, William Ruckelshaus and Lee Thomas, that experience with the agency's workings and mission was the single most important qualification for the Board's chairman.

Boundary Exercises

While the foregoing account helps explain how the SAB has avoided capture by partisan political interests, it fails to answer an equally important question: given the Board's carefully constructed character as a provider of *regulatory* science advice, how has it been able to maintain its scientific authority on the one hand, and on the other to avoid being labeled as captive to EPA's mission? The answer lies to some extent in the very successful boundary work by which the Board and its committees have held themselves aloof from the appearance of making policy.

SAB's reputation for independence can be traced back to its origins in the Office of Research and Development, where its primary duty was to advise the agency on its research programs. This function placed the Board both actually and symbolically at the furthest possible remove from the day-to-day politics of regulation. Advising EPA on research remains one of SAB's most important tasks and partly accounts for both its public image and its self-image as a primarily "scientific" body. EPA and the Board have both worked hard to preserve this image, although even in this relatively apolitical sphere of the advisory process, SAB's activities inescapably blend science with policy. The Board's involvement in drawing up a long-term basic research strategy for the 1990s provides an instructive example.

In the spring of 1987, EPA administrator Lee Thomas charged

SAB's Research Strategies Committee to review EPA's R&D program and to recommend specific types of research to meet the agency's long-term policy objectives. The committee's name implicitly established a boundary; it was to advise EPA on "research," not policy. Its report, issued a year later, emphasized the role of science in detecting and protecting against risk: "research is the most fundamental of the tools that promote environmental quality."[32] But on closer reading the committee's recommendations seemed to be directed as much at EPA's long-range policy options as at its R&D strategies. For example, the committee embedded all of its specific proposals relating to R&D within a conceptual framework favoring "a strategic shift in emphasis from control and clean-up to anticipation and prevention."[33] Strictly speaking, such a recommendation would more appropriately have issued from a policy advisory body such as the Office of Technology Assessment.[34]

It was, however, no random collection of experts that made up the Research Strategies Committee. At its head was Alvin L. Alm, former deputy administrator of EPA, and its five subcommittees included scientists, engineers, and managers with considerable prior experience in environmental decisionmaking. The composition of the ten-member Health Effects Group in particular—it was chaired by Dr. David Rall, director of NIEHS, and its other members were Eula Bingham, Bernard Goldstein, David Hoel, Jerry Hook, Philip Landrigan, Donald Mattison, Frederica Perera, Ellen Silbergeld, and Arthur Upton—showed a (perhaps constructive) bias toward scientists identified with a prohealth policy orientation. The group included current and former agency officials (Bingham, Goldstein, Hoel, Mattison, Rall), researchers critical of the Reagan administration's deregulatory policies (Landrigan, Perera, Upton), and scientists affiliated with public interest groups (Silbergeld).

Recommendations from such a group could hardly have been expected to diverge radically from EPA's known views about areas of needed research. Indeed, many of the agency's familiar biases were subtly in evidence in the subcommittee's report: an emphasis on man-made hazards, a pronounced interest in methods for refining the use of animal studies, and relatively greater concern over air and water pollution than diet. While these biases in no way detracted from the quality or soundness of the subcommittee's recommendations, they may have prevented the group from considering research gaps identified as important by such well-known EPA critics as Bruce Ames of the

University of California at Berkeley. Ames dismisses much of EPA's ongoing research and regulatory effort as misdirected because it downplays big risks such as environmental radon and naturally occurring chemicals ("nature's pesticides") in favor of the relatively small risks of industrial pollution.[35]

The SAB's own contributions to boundary work are particularly in evidence when its advice is closely related to policy. The Board has made conscious efforts to perform its reviews early in the decision-making process, so as to emphasize that it advises EPA only on scientific issues. The Board's reluctance to be drawn too closely into the sphere of policy emerged during an exchange between Norton Nelson and Congressman John Dingell over the role the SAB should play under the 1986 amendments to the Safe Drinking Water Act. A new provision of the law specified that the SAB's comments should be sought "prior to proposal of a maximum contaminant level goal and national primary drinking water regulation."[36] In a letter to Dingell, Nelson made it clear that the Board had no intention of becoming involved in defining EPA's regulatory goals:

> The SAB . . . prefers to focus on issues pertaining to scientific assessment rather than comment on rulemaking, recognizing that the latter involves the weighing of many other factors besides science. In keeping with this approach, the SAB does not plan to evaluate EPA's policy goals when it reviews the technical basis of maximum contaminant levels and primary drinking water regulations.[37]

Nelson's implicit boundary between the "science" of risk assessment and the "policy" of rulemaking clearly served the SAB's institutional purposes well, although, as we have seen, there is little support for the existence of such a boundary in the literature on risk assessment.

SAB's Impact on Policy

While the SAB's prestige within the agency has grown over time, its impact on policy remains surprisingly difficult to measure, largely because of the institutional limitations on the timing and character of the advice that the Board offers to EPA. As with all advisory committee recommendations, SAB's comments on the quality of the scientific support for various EPA policies are not binding on the agency. Moreover, the very success of the Board's rhetorical strategy of distancing itself from policy creates a risk that its advice will be seen as irrelevant to policy. These facts give rise to two generic problems, which may afflict the SAB more severely than committees situated closer to the endpoint of regulatory decisionmaking.

First, since EPA is under no compulsion to go along with the Board's judgment, members are sometimes faced with the frustrating experience of seeing their views ignored or flouted by the agency. In a loosely structured advisory system, obtaining adequate feedback from the agency is always a problem. While EPA administrators or deputy administrators may voluntarily discuss with their scientific advisers the reasons for their final actions, they are not required by law to show such consideration. And even when the agency agrees in principle to be guided by SAB's recommendations, committee members can do little to guarantee that EPA's response will be appropriate or adequate. By the mid-1980s, improved relations between EPA and the SAB had led the administrator to adopt a policy of replying by letter to the Board's reviews; in theory, all committee members who participate in a review were entitled to see the response letter. Nevertheless, one scientist familiar with the SAB suggested that when the Board really wants action it has to look outside the agency: "to talk loudly and make sure there are some reporters in the room."[38]

A second generic problem for SAB arises from the high scientific uncertainty coupled with the high public visibility of many of the questions on which it is required to give advice. As EPA's primary consultative committee, SAB is asked to review the scientific basis for regulatory proposals or methodological approaches with potentially high political impact. Yet, given the unsatisfactory quality of much regulatory science, as well as the scientific community's natural caution about accepting propositions as proven, the tendency for SAB even under these circumstances is to find that the science is not perfect, that doubts and questions remain. As a result, EPA can be left without a scientific go-ahead for making policy in areas of substantial public concern, ranging from permitting field tests for genetically altered organisms to approving marine combustion of hazardous wastes. An ambiguous response from the SAB may produce regulatory paralysis at just those times when the agency is under most intense pressure to act. Strategic framing of issues for the Board offers a possible way out of this dilemma. For example, it is often easier for the Board to comment on the relative merits of different approaches to environmental risk management than to judge whether a particular method is "safe." Traditionally, however, issues have been placed before the SAB case by case and not in "either-or" pairings that might lead to more meaningful recommendations.

The greater the disparity between the volume of evidence and the pressure for action, the less likely it is that the Board can render useful advice. Where the evidence is thin, EPA may be driven to proceed on the basis of pure supposition or on the strength of questionable risk assessments produced by massaging poor data. Asking the SAB to review such exercises exposes both the agency and its advisers to possible embarrassment. The reviewing scientists may follow one of two equally unpalatable courses. They may publicly criticize the agency for acting on the basis of inadequate science, thus damaging EPA's technical credibility. Alternatively, they may be drawn into evaluating EPA's policy choices and making recommendations that go well beyond their own areas of scientific competence. Some experts see such a danger in the area of hazardous waste management, where public concern has forced EPA to act even when the scientific community views the evidence of risk as too weak to justify policy initiatives.

Conclusion

To summarize, since its establishment in the mid-1970s the SAB has emerged as a powerful and respected adjunct to EPA's regulatory programs. Although the NAS and some of the early critics of EPA's science foresaw an expanded role for the Board, it took years of internal disarray and a continuing loss of scientific credibility to force the agency to rethink its relationship with the SAB. The housecleaning that followed the Gorsuch administration, in particular, led to tangible improvements in SAB's status, such as a larger budget and sharply expanded workload, as well as greater willingness on the part of Congress to commit new responsibilities to the Board. By the mid-1980s, EPA staff and SAB members agreed that there was a new spirit of cooperation between the agency and the advisory panel, symbolized by frequent consultation between the EPA administrator and the SAB chairman and by improved feedback from the agency to the advisory committee.

One reason for this apparent success story seems to be an accident of history. After the political debacles of the early 1980s, the reinstatement of a credible scientific advisory system became a high priority for the agency and such gestures as opening up the appointment process publicly reaffirmed the agency's commitment to securing advice in a politically untainted manner. Another reason is the involvement of a representative cross-section of the scientific community—including scientists affiliated with special interest groups—in SAB's extended advisory network. The practice of drawing on a diversity of expert perspectives reduced the risk that any particular group would complain of bias or underrepresentation. Perhaps the most important reason, however, is that the boundary drawn by the SAB between its work (labeled "science") and EPA's activities (labeled "policy") has insulated the Board from overly close identification with EPA's regulatory goals and related implementation strategies.

But some of the factors that account for the SAB's success—its wholly advisory status, its self-defined distance from policy decisions—also make it difficult to evaluate its impact. The SAB's advice to EPA is generally embedded in a complex procedural matrix, which often includes consultation with other, more focused advisory bodies. Accordingly, to get a sense of how much legitimacy EPA derives from the SAB, one must look at the Board's contribution in context, as one of several significant inputs into a rulemaking proceeding. Two such studies are undertaken in Chapter 9, which analyzes SAB's impact, and that of expert advice more generally, on EPA's guidelines for cancer risk assessment and its regulation of formaldehyde.

6

The Science and Policy of Clean Air

The institutionalization of the Clean Air Scientific Advisory Committee (CASAC) into EPA's decisionmaking on air pollution control offers in some ways an even more dramatic example of mutual accommodation between scientists and policymakers than the experiences of the Science Advisory Board. Created by statute to remedy perceived deficiencies in EPA's use of science, CASAC has moved from an uncomfortable and highly critical relationship with the agency to one that both parties regard as cooperative and productive. This evolution is especially noteworthy in view of the fact that CASAC provides a model for other SAB committees engaged in reviewing the technical basis of EPA standards.

The complex negotiations through which CASAC and EPA resolved their differences are documented in the written record of interactions between scientists and agency staff during the first and second revisions of the ozone standard. The nesting of scientific review within a wider context of political conflict, marked by litigation and procedural wrangling, at first complicated EPA's efforts to reach an accommodation with the advisory committee. The protagonists eventually reached agreement, however, on how to deal with matters over which both wished to assert control. Boundary drawing figured importantly in the exercise: the basic boundary between science and policy had to be carefully readjusted so that neither EPA nor CASAC could claim exclusive jurisdiction over key elements of the standard-setting process. CASAC's involvement in another proceeding, the revision of the carbon monoxide (CO) standard, illustrates some of the benefits that EPA has gained from enlisting the committee as an ally rather than confronting it as an adversary. But the case also indicates that there are limits on any advisory committee's capacity to assure the quality of the basic science used by EPA.

CASAC and the NAAQS Process

Few of EPA's standard-setting obligations entail as complex and protracted an elaboration of regulatory science as the five-year review of National Ambient Air Quality Standards (NAAQS) for criteria pollutants mandated by the 1977 amendments to the Clean Air Act. As described in Table 6.1,[1] the review process begins with a thorough search and analysis of the scientific literature, the findings of which are condensed into a "criteria document" by EPA's Environmental and Criteria Assessment Office (ECAO).[2] The criteria document is intended to "reflect the latest scientific knowledge" useful in identifying all effects on health and welfare caused by pollutants in the ambient air, and review by experts is designed to guarantee its soundness.[3]

The practice of having air quality criteria documents reviewed by outside experts was established even before the 1970 amendments to the Clean Air Act. The first such document, dealing with sulfur oxides, was issued in March 1967 by the National Air Pollution Control Administration (NAPCA) of the Department of Health, Education and Welfare; it underwent two years of review before being

Table 6.1. The NAAQS review process.

EPA conducts a search of all known scientific literature* concerning the criteria pollutant and consolidates the relevant studies into a criteria document that becomes the scientific basis for the standard. The criteria document is circulated to selected researchers and made available for public comment.
EPA develops exposure and sensitivity analyses that compare the relationships between concentrations of the pollutants in the air and the resultant health effects and tests various assumptions regarding the comparisons.
EPA drafts a staff paper that, among other things, evaluates the key studies in the criteria document and identifies the critical elements to be considered in the revision of the standards.
CASAC reviews the criteria document and staff paper and drafts a letter of closure advising the EPA administrator on the scientific adequacy of the documents.
EPA prepares a proposed Federal Register package that is reviewed at a senior level within EPA before being sent to the Administrator for a final decision.

*As a matter of policy, EPA includes in the criteria document only research that has been published by a professional scientific journal. Accordingly, research relied on by EPA has already been subjected to normal prepublication peer review.

reissued in final form in 1969.[4] The review process was managed by the National Air Quality Criteria Committee, a broadly representative body with members from industry, universities, conservation interests, and all levels of government. Subsequently, the same procedures were followed for four additional pollutants: particulate matter, carbon monoxide (CO), hydrocarbons, and photochemical oxidants (later designated ozone). These documents became the basis for the NAAQS promulgated by EPA in 1971.

This approach, however, struck EPA's critics as insufficiently rigorous. During deliberations on the 1977 Clean Air Act amendments, the House Committee on Interstate and Foreign Commerce noted that NAS and NIEHS, among others, viewed the scientific basis for the NAAQS as inadequate. The committee's own conclusion was that the standards established by EPA were, if anything, too lenient. In view of the chronic deficiency of data and the controversial character of the standards, the committee determined that EPA should be required to perform regular reappraisals of the NAAQS, accompanied by more credible scientific review. Accordingly, the 1977 amendments established a new advisory committee, CASAC, to serve as an "independent source of review and advice" to EPA. Congress sought to ensure CASAC's independence by placing restrictions on its membership. The law stipulated that three out of CASAC's seven members should be, respectively, a physician, a representative of NAS, and a representative of a state air pollution control agency. Both real and apparent conflicts of interest were to be avoided among individuals appointed to CASAC. Finally, unlike the earlier National Air Quality Criteria Committee, CASAC was to select its members solely "on the basis of their special expertise in the fields of environmental toxicology, epidemiology and/or clinical medicine, or in the fields of environmental or ecological systems."[5]

CASAC intervenes in the standard-setting process at several different points, beginning with ECAO's release of the "first external review draft" of the criteria document.[6] Issues addressed by the agency staff and the committee at this stage include the design and conduct of the studies analyzed in the criteria document, the use of statistical techniques, and the degree of consistency or lack thereof among the studies surveyed. CASAC also considers questions of organization and presentation and advises the agency on ways in which the document can be made more accessible to users outside the agency. The committee's suggestions and comments from the general public provide the basis for preparing a second review draft, which is resubmitted to CASAC. The process continues until the point of "final closure" at which the committee concludes as a body that it has no further substantive criticism of the document.

The second major stage in the review process commences with the preparation of a "staff paper" by the Office of Air Quality Planning and Standards (OAQPS). Explicitly designed to bridge the gap between science and policy, the staff paper lays the basis for the judgments that the administrator must make in setting ambient standards.[7] One objective of the staff paper, therefore, is to identify the key studies that are relevant to determining a pollutant's health effects and the possible numerical levels that will protect the public. In other words, the staff paper deals with the implications of the raw data and studies summarized in the criteria document. These interpretations, which necessarily include elements of science policy, are reviewed by CASAC at an open meeting; as with the criteria document, closure must be reached with the committee before EPA proceeds with the development of a standard. This is a key stage of the rulemaking process, for it enables the agency and its expert advisers to resolve any possible disagreements over the interpretation of evidence.

Although the review of criteria documents falls in the gray zone between science and policy, CASAC continues to view its own role as that of purveying purely scientific advice. This is reflected in the way the committee typically frames its findings. For example, in its letter to the administrator on two documents related to EPA's proposed air quality standard for CO, CASAC found both to be "scientifically balanced and defensible" summaries of current knowledge.[8] CASAC also advises EPA on the direction of its air pollution research programs. Thus, in 1983–84 CASAC formed two subcommittees to address the agency's research needs on the health and welfare effects of criteria pollutants and approved the formation of a subcommittee to study the standard-setting process and recommend means of improvement. To understand how the committee maintains its independent stance, however, one has to look in detail at its actual involvement in standard-setting.

Science and Standards

The promulgation of a new ozone standard in 1979, representing EPA's first complete reanalysis of one of the original NAAQS issued in

1971, lives in the agency's institutional memory as an example of how not to use scientific advice. In explaining this assessment, EPA insiders point to the patent lack of harmony that prevailed between the agency and its advisory committee through much of the revision process. Outsiders, however, have read the events in a different way, blaming the problems of the 1979 rulemaking more on EPA's wrongheaded legal, political, and economic judgments than on its use of science. If the latter interpretation is correct, it supports Collingridge and Reeve's argument that political factors nullify the power of science to make a positive difference in standard-setting.

An authoritative exposition of the view that science was subservient to politics in the ozone NAAQS process can be found in R. Shep Melnick's study of the implementation of the Clean Air Act.[9] Melnick argued that the scientific evidence available to the agency would have supported a standard anywhere between 0.08 parts per million (ppm), the existing level, and 0.25 ppm. EPA's decision to settle for 0.12 ppm was dictated, in Melnick's view, not by the evidence pertaining to health effects, but by a complicated interplay of political interests inside and outside the agency. Pressures for relaxing the standard included the threat of legal action by the American Petroleum Institute (API), the prospect of new legislation mandating periodic review of the NAAQS, and arguments by White House economic advisers that the existing standard was unnecessarily costly. Within the agency, a split developed between the "strict constructionists" of the Clean Air Act and those favoring a more flexible interpretation. The former insisted that the primary NAAQS for a criteria pollutant should be established on the basis of health risks alone, and that this, in turn, meant pegging the standard at a level below which there was no evidence of harmful effects. The moderates in the air program opted instead for a level of control that would permit the agency to ignore rural areas and to redirect its enforcement efforts toward the more pressing problem of urban pollution.

Positions within the agency ultimately hardened around a final standard of 0.12 ppm, below which the costs of control were felt to be disproportionate to benefits. Melnick has argued that even this numerical level represented a face-saving effort by the agency. EPA had insisted loud and long on the validity of studies showing effects at 0.15 ppm and it could not abandon this posture without losing credibility. The 0.12 standard offered a margin of safety over 0.15 ppm, as required by the Act.[10] But EPA's reading of the evidence cannot be understood in isolation from the positions of other major players in the proceedings, all of whom took advantage of scientific uncertainty to construct their own assessments of the risks of ozone.

Procedural Issues

A threshold issue at stake in the ozone proceedings was EPA's freedom to choose who would review its scientific determinations. Early in 1977, EPA formally declared its intention to revise the ozone standard and requested SAB to assist it in the development of a new criteria document.[11] Since the CASAC review mechanism was not yet officially in place, SAB created a Subcommittee on Scientific Criteria for Photochemical Oxidants, chaired by Dr. James L. Whittenberger, former dean of the Harvard School of Public Health, and consisting of nine additional members.[12] The subcommittee reviewed three drafts of the criteria document; complete drafts were considered in November 1977 and February 1978, and a draft of chapter 1 was reviewed in March 1978. Each time, the panel unanimously concluded that the document was scientifically unacceptable and roundly criticized the agency for failing to revise it in accordance with the panel's previous comments and suggestions.

Concurrently with these proceedings, EPA in early 1977 appointed a separate panel of consultants to provide additional advice on the health effects of ozone. Dr. Carl Shy, who headed the panel (known accordingly as the Shy Panel) and selected its remaining members, was widely regarded as an advocate of keeping the ozone standard at the then current level of 0.08 ppm. A former EPA scientist and a consultant for the American Lung Association, Shy had appeared before a congressional committee to defend the existing standard just weeks before he was called upon to organize an advisory panel for EPA.[13] Thus, no one was surprised when his panel submitted a report to the agency endorsing the 0.08 ppm standard. SAB and API both objected to EPA's soliciting advice outside what they regarded as the normal channels of scientific review. Their charges of irregularity gained substance when EPA tentatively decided not to relax the 0.08 ppm standard even before completing the second draft of the criteria document. In an effort to deflect criticism, EPA entered the Shy Panel's report into the docket and agreed to accept comments on it. But the charge that the agency had unlawfully consulted with a biased panel remained alive throughout the ozone rulemaking and was eventually incorporated into API's lawsuit against EPA.

Controversy also arose over the appropriate timing and scope of review by the SAB. EPA, as already noted, presented neither its final ozone standard nor the underlying criteria document to SAB for approval. API strenuously objected to this course of action, arguing that it resulted in EPA's use of highly questionable science as a basis for the final rule. In particular, EPA relied on four items that API regarded as illegitimate bases for standard-setting: (1) a table of health studies that EPA itself admitted "must be used with caution," (2) the report of the Shy Panel, (3) the results of a disputed risk-assessment exercise, and (4) an obscure scientific study that was neither cited in the original regulatory proposal nor discussed by the SAB.[14] While none of these studies or reports could be characterized as "peer-reviewed science," EPA took the position that it was not essential for SAB to approve any of them, since the Board's role was purely advisory. In a nice display of boundary work, EPA insisted that the development of a criteria document was a separate activity from standard-setting. SAB's involvement, the agency maintained, was required for the former but not for the latter.

To some, however, this view of the NAAQS standard-setting process appeared crabbed and simplistic. It was clear that restricting SAB (or CASAC) review to the criteria document alone left enormous discretion for scientific decisionmaking in the agency's hands.[15] A criteria document merely surveys all the existing research on the effects of a pollutant. It is a massive undertaking—the revised ozone criteria document prepared by EPA between 1984 and 1986 consisted of five volumes and was hundreds of pages long—and it is only the first step in a complex analytical process. To set a standard, the agency has to extract from this scientifically overloaded record the few key studies and analyses that bear most significantly on public health protection. A strict separation between criteria-document review and standard-setting would permit this selection process, and the subsequent assessment of health risks, to proceed without any supervision by nonagency scientists. Indeed, as in the case of the 0.12 ppm ozone standard, the agency could use its discretionary powers to select studies and "expert" opinions that were rejected by its scientific advisers. To prevent such abuses of discretion, Lester Lave and Gilbert Omenn proposed that the scientific review of standard-setting should be revised to require another round of scrutiny after EPA decides which studies will actually be used as the basis for action.[16] This is the approach the agency currently follows. As will be seen below, this procedural change served an important consensus-building function in the agency's later decisions with respect to ozone.

Disputes over Data

EPA and the SAB subcommittee disagreed both about the meaning of individual studies and about the conclusions to be drawn from the evidence as a whole. One dispute centered on the scientific legitimacy of a study by A. J. DeLucia and W. C. Adams of ozone's adverse impact on pulmonary function.[17] The SAB scientists contended that EPA could not fairly rely on this study because it contained serious methodological flaws.[18] First, contrary to EPA's initial claim, DeLucia and Adams had not found statistically significant adverse effects in subjects exposed to 0.15 ppm of ozone. Second, the SAB reviewers questioned the validity of using heavily exercising healthy subjects to approximate the impact of ozone on specially sensitive populations, such as asthmatics. Third, SAB members were not satisfied that subjective symptoms of discomfort (chest pain, congestion, cough) amounted to evidence of adverse health effects. Fourth, the reviewers suggested that the use of mouthpieces to dispense ozone to study subjects provided so inaccurate a measure of actual exposure as to invalidate the study. In its final rulemaking on ozone, EPA rejected some of these criticisms, but conceded that there was no statistically significant evidence of health impairment at levels below 0.30 ppm.[19] Nonetheless, the agency continued to accept the DeLucia and Adams study as establishing a *qualitative* likelihood of adverse effects at lower exposures, and hence as appropriate for use in standard-setting.

To the student of social construction in science, these disagreements look like a classic case of "experimenters' regress." With no previously shared understandings about what constitutes an "adverse health effect" of ozone—let alone how to measure such effects—the agency and its advisers naturally fell to arguing over particular features of experimental design, such as the selection of subjects. Any attempt to resolve the debate was complicated by the fact that it was embedded in a political context that permitted disputes over evidence to be redefined as disputes over turf and territory. EPA tried to win the turf battle by claiming, through a familiar species of boundary work, that the agency and not the reviewers had the right to determine how the data on health effects should be interpreted. Specifically, both in the ozone rulemaking and in the partially concurrent proceedings on lead, EPA claimed that the determination of adverse health effects was a question of science policy falling within the sphere of administrative discretion. In both cases, EPA used its self-proclaimed prerogative to label as an "adverse health effect" changes that were much less severe than clinical disease symptoms or other permanent physical impairments. Supported neither by a medical or scientific consensus nor by a negotiated political accommodation, these determinations were bound to prove vulnerable to charges of illegitimacy.

The second major disagreement between EPA and SAB concerned the agency's use of a nonvalidated procedure for assessing the risks of ozone. This decision-analytic technique (also called a risk-assessment technique by EPA) involved a sampling of expert opinion on the probable effects of ozone on sensitive individuals.[20] The agency convened a panel of nine physicians, questioned them about the health effects of ozone, and averaged their responses to get a usable measure of risk. While the SAB applauded EPA's efforts to develop a systematic approach to risk assessment, committee members were by no means persuaded that the methodology developed by the agency's experts, Thomas Feagans and William Biller, was technically sound.[21] In its lawsuit against EPA, API was much more vehement:

> This technique is simply a means of polling undisclosed "experts" for their opinions regarding the chance that a hypothetical person will experience effects from exposure to various concentrations of ozone. Its effect is essentially to repackage opinion evidence in a form to make it appear more reliable and precise.[22]

Although API's characterization may have been unduly harsh,[23] it induced EPA to back away from the Feagans-Biller approach, acknowledging that more study was required before the method could validly be incorporated into rulemaking.

The White House and the Environmentalists

Challenges to EPA's reading of the scientific data on ozone came not only from SAB and the petroleum industry, but also from White House economic analysts concerned about the cost of the final ozone standard. A literal reading of the Clean Air Act seemed to rule out the consideration of economic effects in setting primary standards protective of public health. In cases arising under similar provisions of the Clean Water Act, federal courts had held that standards should be set solely on the basis of health effects and not economic considerations.[24] The Council on Wage and Price Stability (COWPS), however, pressed the view that economic costs should be taken into account when the scientific evidence was uncertain. In comments to the Regulatory Analysis Review Group (RARG), COWPS recommended that the standard be weakened by 100–150 percent from the levels established in 1971, a position the economists felt was consistent with the Act's protective mission.

Congressional hearings on the legitimacy of executive branch review of environmental regulations provided a forum for President Carter's chief economic advisers, Charles Schultze and Alfred Kahn, to defend their proposal. Schultze, in particular, noted that "the scientific evidence more and more doesn't suggest some specific threshold of ambient concentration below which there are no harmful effect[s] and above which there are. Usually there is a continuum of effect going all the way down to zero."[25] Given the uncertainties involved in identifying potentially harmful concentrations and establishing margins of safety, Schultze argued, it was simply a matter of "prudent public health judgment" not to set standards any more stringently than necessary. In other words, given a range of scientifically supportable values for the ambient standard, it was appropriate for the agency to select the standard that was most prudent from the standpoint of both health effects and economic impact. According to COWPS's (presumably not peer-reviewed) analysis of the ozone data, there was no evidence that even a 0.16 ppm standard would be significantly less protective of health than the 0.08 ppm standard established in 1971.[26] Since the costs of compliance seemed to rise sharply for stricter levels of control, the White House economists saw no reason why EPA should not be pressured to choose a standard somewhere in the range of 0.16 ppm.

Two leading environmental groups, the Natural Resources Defense Council (NRDC) and the Environmental Defense Fund (EDF), sharply contested this view of EPA's responsibilities on scientific as well as legal grounds. Both were prepared to acknowledge the existence of uncertainty, but they argued that EPA's choice of data from a disputed record was strictly constrained by the governing statute.[27] The environmental advocates stressed the agency's obligation to make "precautionary" policy and cited with approval the judicial doctrines from *Ethyl Corp.* and other cases that instructed EPA to accept a lowered standard of proof where public health was at stake. They dismissed COWPS's reading of the scientific evidence as fundamentally flawed because it failed to recognize EPA's duty to put health ahead of economics in setting the NAAQS.

Given their presumption that costs were irrelevant to the primary ozone standard, the environmental groups were naturally suspicious of all contacts between EPA and the executive branch economists. Their fears were confirmed when EPA eventually moved to a 0.12 ppm standard rather than the 0.08 and 0.10 ppm levels previously under consideration. The fact that this move followed intensive consultations between the EPA administrator and senior presidential advisers like Kahn and Schultze provided the environmentalists with prima facie evidence that the agency had given unlawful primacy to economic considerations.[28] That so many of these discussions were held after the record on ozone was formally closed gave the proceedings an added air of secrecy and illegitimacy. The legality of postclosure contacts with White House personnel, in particular, became an important issue in subsequent litigation over the ozone standard.

API v. Costle

EPA's final action on the ozone standard satisfied no one completely. Both API and NRDC sued the agency; API claimed that there was no scientific support for the 0.12 ppm standard, and both parties argued that the standard was invalidated by procedural irregularities. API also challenged the legality of EPA's decision not to seek SAB's approval of the final criteria document and standard. In addition, API argued that the Shy Panel was an advisory committee within the meaning of FACA and that EPA had failed to comply with FACA requirements in convening and relying on these experts. For its part, NRDC argued that the contacts between EPA and the White House after the comment period were unlawful. NRDC also objected to EPA's last-minute decision to place in the record a staff paper which the agency claimed was the primary basis for the change from the 0.08 ppm standard to the 0.12 ppm standard. This action, taken only a day before promulgation of the final rule, precluded any meaningful opportunity for the public to comment on a crucial part of the rationale for the final standard.

The D.C. Circuit Court disposed of most of these objections in relatively short order. Although the court admitted the seriousness of NRDC's procedural claims, it held that these issues could not be raised for the first time on appeal.[29] With regard to the Shy Panel, the court did not find it necessary to consider the possible application of FACA, since there was no evidence that EPA had relied on the panel for its final standard.[30] Indeed, the agency's action appeared to go against the Shy Panel's recommendation that the standard should not be relaxed from the 0.08 ppm level. Failure to submit the standards to the SAB was a more substantial problem because it appeared to violate EPA's procedural obligations under ERDDA. Nonetheless, the court considered this failure insufficient by itself to invalidate a standard that otherwise appeared to be adequately supported by the record.[31] The decision thus upheld EPA's scientific and legal judgments on ozone, though it sent a clear message that the court did not look kindly on the agency's undisciplined management of scientific advice.

Redefining CASAC's Role

Given the foregoing history, EPA obviously had a substantial stake in restoring its scientific credibility during the second comprehensive review of the basis for the ozone standard, beginning in 1984. The criteria document developed by the agency was discussed by CASAC at a three-day meeting in March 1985, and the transcripts of that meeting provide evidence of a newly cooperative attitude by agency staff toward their scientific advisers. Relations between CASAC and EPA appeared to be significantly smoother than those between EPA and SAB during the 1978 ozone proceedings. The primary reason for this change was the shift from an adversarial to a negotiating mode in the consultations between the agency and its advisory committee. The issues confronting EPA in 1986 were not very different from those that gave rise to such bitter conflicts in 1978. However, the process for addressing these issues through the advisory system had changed dramatically, creating more opportunities for dialogue between agency officials and CASAC, particularly with regard to issues of science policy.

In its official statement concerning the scope of the new ozone criteria document, EPA seemed to redraw the same boundaries on which it had insisted at the earlier round of revision. The criteria document, EPA said, addressed only the scientific analyses required to establish ozone's effects on health and welfare.[32] Control techniques and control strategies would not be discussed. Nor would the document address three additional sets of issues germane to standard-setting: (1) determination of what constitutes an adverse effect, (2) assessment of risk, and (3) determination of a margin of safety. These reservations reflected the agency's judgment that "while scientific data contribute significantly to decisions regarding these issues, their resolution cannot be achieved solely on the basis of experimentally acquired information."[33] In other words, EPA again seemed to be characterizing these issues as science policy, and hence as inappropriate for inclusion in a fundamentally scientific review document, although they might eventually be brought before CASAC during the review of the staff paper. In practice, however, most of these issues were extensively discussed even in the phase of criteria-document review, showing that EPA was prepared to grant CASAC substantial say in matters on the boundary of science and policy. The agency's flexibility was perhaps most apparent in its exchanges with CASAC over the definition of adverse health effects.

Adverse Effects

The question whether particular pollutant-induced physiological effects should be viewed as "adverse" to health has long been a focus of controversy under the Clean Air Act. The point was thoroughly debated, for instance, during EPA's attempts to control the lead content of fuels[34] and to establish the NAAQS for lead. A challenge by the Lead Industries Association (LIA) to the final ambient lead standard proposed by EPA squarely presented the issue of what effects may be considered "harmful" under the Act. LIA maintained that by setting the standard so as to protect the public against subclinical effects the EPA administrator had exceeded his statutory authority and had taken action that was more stringent than Congress mandated. In upholding EPA's regulatory decision the D.C. Circuit Court essentially ratified the administrator's contention that this was a valid discretionary judgment:

> [T]he precautionary nature of the statutory mandate to protect the public health, the broad discretion Congress gave him to decide what effects to protect against, and the uncertainty that must be part of any attempt to determine the health effects of air pollution, are all extremely difficult to reconcile with LIA's suggestion that he can only

> set standards which are designed to protect against effects which are *known to be clearly harmful to health*.[35]

Pursuing this line of reasoning, the court rejected LIA's suggestion that there should be a medical consensus as to whether the effects on which standards are based are genuinely harmful. Such consensus, the court felt, would be impossible to achieve, given the uncertainties involved in measuring the effects of air pollutants; medical disagreements would be sure to arise in carrying out such measurements.[36]

What the court overlooked, however, was the scientific community's probable reaction to a decision of this kind. At the same time that the court was bowing to the science policy paradigm and expanding the administrative agency's discretion, EPA's scientific advisers were drawing their own boundaries around the issue of "adverseness," boundaries that emphasized the "science" involved in this determination. The controversies during the first ozone NAAQS revision showed that EPA could not, as a political matter, ignore its advisers' perception that scientists had to be involved in the definition of adverse health effects. The agency's decisionmaking strategy in the second ozone proceeding recognized the right of CASAC members to comment on this issue.

During the March 1985 review of the ozone criteria document, several members of CASAC agreed that scientific judgment had to play a part in determining adverse effects,[37] but they worried that this exercise might lead the committee into regions lying beyond their spheres of competence. Dr. James Crapo articulated both sets of concerns:

> . . . once we start defining these small changes, whether they be in an FEV–1 or a biochemical change in the lung, the question of whether that represents an adverse effect and how adverse it is, I don't know of any body of people that are better able to resolve that question accurately than a group of people that derive the data and know its weaknesses and strengths . . .
>
> Our weakness is that we don't understand the political/economic issues as well, perhaps, as we should for making those kinds of assessments so I don't see that it's the place for this committee to try and make the final decision as to where the regulation ought to occur.[38]

EPA's response to this soul-searching was highly conciliatory. Speaking for the agency, Lester Grant indicated that EPA was prepared to discuss its own views about adverse effects with the committee either in the course of the criteria-document review or in the subsequent staff paper prepared by OAQPS.[39] Dorothy Patton from EPA's Office of General Counsel was quite explicit about CASAC's right to comment on issues with a possible policy dimension:

> I hope nothing that you have read in the criteria document or that anyone has said leads you to think that you are not free to give your scientific opinion, your medical/scientific/biological opinions on adverse effects. Now, if . . . the Administrator doesn't agree with you on any of these things, he has to explain the reasons for his differences, his different views. So, again, the main thing I wanted to say was that you are free to comment to EPA on anything that you feel like commenting about.[40]

In other words, EPA recognized that CASAC had an independent right to comment on any aspect of a science policy issue that it saw as scientifically relevant. If the agency disagreed with CASAC's analysis on policy grounds, then it would simply have to explain its own position in public.

The committee, for its part, was also careful not to assert exclusive jurisdiction over adverse effects. On the third day of the meeting Chairman Lippmann summed up his colleagues' position, noting that "clearly there has been a consensus view of the committee that anything that contributes substantially to the development of chronic disease would be considered an adverse effect."[41] Lippmann emphasized, however, that the committee was in no position to provide EPA with a "magic formula"[42] for determining adverse effects, and that this decision could not always be made as a matter of scientific judgment. Accordingly, decisions as to where CASAC's role should end and where the committee should "abdicate to the political level"[43] would have to be made on a case-by-case basis.

CASAC's review of the draft ozone staff paper in April 1986 gave the committee further opportunity to discuss adverse effects. Curiously, the expansion of the committee's scope of review led to an extremely precautionary set of instructions to EPA. CASAC found that a range of common ozone reactions, including short-term lung function disability, coughing, and eye irritation, were "adverse" and recommended that EPA tighten the range of regulatory options it was considering in light of this finding.[44] In particular, the committee saw no scientific support for relaxing the existing 0.12 ppm standard, although OAQPS included a proposed standard of 0.14 ppm in its range of options. Indeed, the committee noted that even the existing standard did not appear to provide the adequate margin of safety required by the statute.

Risk Assessment

The controversy over EPA's unvalidated method of risk assessment during the first ozone review alerted the agency to the need for a more rigorous approach to accounting for subjective expert judgments about health risks. To placate its critics, EPA established a special SAB subcommittee on Health Risk Assessment, sought advice from professional decision analysts, and, finally, contracted with Thomas Wallsten of the University of North Carolina to review the literature on "expert elicitation" (also known as "judgmental probability encoding"), a set of techniques for quantifying the variations in expert opinion about risk.[45] These initiatives led to the adoption of a method that the agency felt it could use with much greater confidence than the earlier Feagans-Biller approach. One measure of this change in perspective was EPA's decision to incorporate expert elicitation into the development of the revised ambient standard for lead,[46] a strategy that drew favorable reactions from CASAC.

But EPA still had considerable resistance to overcome before generalizing the use of expert elicitation to other regulatory proceedings. At the March 1985 CASAC meeting, for instance, Dr. James Whittenberger, chairman of the SAB committee that had reviewed EPA's ozone criteria document in 1978, recalled with pique EPA's previous flirtation with such a procedure:

> Subsequent to the subcommittee review of the document, a number of events took place in OAQPS which amounted to an independent review of the criteria document that was not under the auspices of the SAB, and the development of a risk assessment methodology which had not been previously tried or validated and which was subject to a great deal of subsequent criticism . . . The standard eventually was changed, based largely on a modified delphi approach rather than on any well established set of procedures that the Agency had.[47]

More generally, Whittenberger strongly disapproved of a legal regime that gave EPA final authority to determine such core science policy issues as the methodology of risk assessment:

> For example, on page 67 [of the criteria document], it states that . . . the final identification of those effects that are considered

adverse and the final identification of at risk populations are both the domain of the administrator. I feel like Alice in Wonderland when I read statements like that. I don't care what the law says. It's just absurd for the people who prepared this criteria document and critiqued it not to be the ones to do the risk assessment and the ones that determine what an adverse effect is.[48]

It appears, however, that Whittenberger's concerns were misplaced, for EPA's response was again conciliatory. Agency representatives indicated that their risk-assessment models would be available for review by the committee along with the staff paper.[49] Harvey Richmond, the agency's project officer for risk assessment, interacted with the committee during the March CASAC meeting and helped allay skepticism about the agency's unconventional approach to placing quantitative values on subjective expert judgments.[50]

The committee's deliberations evidently persuaded CASAC member Dr. James Ware that it was reasonable to speak of two approaches to risk assessment when dealing with problems of air pollution: the traditional method of "statistical risk assessment," a quantified, data analytic method for combining results from a variety of studies, and the subsidiary technique of expert elicitation, which should be employed only when the data are inadequate for conventional statistical analysis.[51] Other committee members were initially unconvinced that the second approach could ever be meaningfully used. Dr. Timothy Larson commented, for example:

I guess I suffer a bit from not being too familiar with the details, but the expert elicitation process in this case bothers me a lot . . . What is the mean of what ten people believe to be adverse? It doesn't necessarily mean that that's what is real or that's what it is. It's an average of a set of expert's opinions about things that they perhaps have not such great confidence about.[52]

Chairman Lippmann, who was familiar with the agency's ongoing efforts to use expert elicitation in the lead case, also expressed misgivings about the technique's rigorousness.[53] But Ware was more optimistic, arguing that the two approaches could usefully supplement each other. First, the statistical approach could be used to integrate the data as far as possible, narrowing the range of uncertainty; then, expert elicitation could be used to bridge any further gaps in the science.[54] Ware noted that a committee like CASAC provides one mechanism for organizing the scientific community's subjective reactions to objective information; this, he felt, was no different in essence from what EPA was trying to achieve through expert elicitation.[55]

Discussions between agency experts and CASAC over the validity of expert elicitation left many questions unanswered, even after the method was more fully elaborated in the ozone staff paper. Is the technique capable of yielding consistent responses? Can the agency manipulate the answers through the selection of experts? Does the quantification of essentially subjective judgments convey a false sense of precision about risk estimates? By casting the debate in artificially technical terms, is the process likely to transfer an impermissible degree of control from nonexperts to experts?[56] Because of these questions, even those EPA advisers who believe that expert elicitation yields a meaningful measure of the uncertainty associated with particular expert opinions are unsure whether individual elicitations should ever be combined in order to simulate a collective expert judgment.

The review of the staff paper could not and did not definitively resolve these issues. But it did illustrate two ways in which advisers can assist agency decisionmakers operating at the margins of scientific certainty. The first contribution made by CASAC was to prevent an overextension of the expert-elicitation approach by the agency and its expert consultants. Committee members pointed out that it was inappropriate to use this method to develop a dose-response function for human health effects. The information on health effects was sufficiently strong and consistent to permit ordinary statistical analysis. Expert judgment was far more necessary in extrapolating from the animal data to humans, and this was therefore a more appropriate area for encoding expert judgments.[57] The second beneficial consequence was to persuade a growing number of CASAC members that the agency's approach was indeed promising enough to warrant inclusion in EPA's array of formal decisionmaking techniques. By ensuring that it would have the support of these experts in future debates about risk assessment, the agency created a defense against the charges of scientific isolation and incompetence that had marred its earlier rulemaking on ozone.

Specially Sensitive Populations

EPA indicated at the March 1985 CASAC meeting that the appropriate margin of safety for the ozone standard would be addressed in the staff paper.[58] However, another issue relevant to the precautionary character of the standard was treated in the criteria document and discussed by CASAC: the definition of specially sensitive (or "at risk") populations. EPA proposed to designate three groups of individuals as particularly at risk: persons with preexisting disease, "responders" (that is, individuals who for medically unexplained reasons are exceptionally sensitive to ozone), and heavy exercisers.[59]

Industry representatives present at the three-day criteria-document review were unimpressed by EPA's approach. The spokesman for the Chemical Manufacturers Association objected to the "essentially vague" characterization of the responders and the subgroups with preexisting disease.[60] He suggested that it was inappropriate for EPA to designate certain subpopulations as sensitive unless the agency could provide medically or biologically valid criteria for distinguishing them from the general population. A representative of API, Dr. Eugene Holt, made a more pointed attack on the scientific basis for EPA's statements with respect to asthmatics and persons with preexisting disease.[61] In his view, a more thorough analysis of deficiencies in the studies EPA used in identifying these groups would have disclosed their unreliability. Indeed, Holt suggested that these studies should be reconsidered at another public peer-review session.[62] Both industry experts, in other words, stressed the need for more hard science in the designation of specially sensitive populations and were reluctant to recognize such groups as significant in the absence of stronger indicators based on biological or epidemiological information.

By contrast, CASAC seemed content with EPA's scientifically "soft" approach to classifying specially sensitive populations. One member expressed uneasiness at treating "responders" as a distinct subgroup in the absence of any physiologically differentiating factors other than the response to ozone.[63] Other committee members noted, however, that it was important to provide evidence of a heterogeneity of response to the pollutant, even if variations among individuals could not be explained in biological mechanistic terms. Such evidence, it was felt, would be significant for purposes of risk assessment and would also provide guidance to the administrator in making his final policy decision.[64] For the rest, the committee agreed that further research was warranted to determine why a substantial subgroup of the population exhibits unusual sensitivity to ozone, and that this issue should be considered in defining EPA's future research program on the pollutant.

The Carbon Monoxide Controversy

CASAC's involvement in the revision of the carbon monoxide standard provides further evidence of the ways in which a properly functioning advisory process can insulate regulatory agencies against attacks based on their use of science. But the CO study also indicates that there are aspects of scientific legitimation—particularly, quality assurance in the production of basic scientific knowledge—that advisory committees are not institutionally well equipped to deliver.

EPA began revising the 1971 CO standard in late 1978. Five years later, when the agency had considered the alternatives and was on the point of issuing a revised standard, the *Washington Post* disclosed that Dr. Wilbert S. Aronow, a scientist at the Veterans Administration (VA) who had carried out most of the research supporting the new standard, had been investigated by the VA and FDA for alleged falsification of research results. This revelation threatened to undo all of EPA's progress toward a new standard. Scientific support for the original 1971 standard was generally conceded to be weak.[65] The key study supporting that standard was widely dismissed as unreliable both because it was based on subjective responses from patients and because later researchers had not been able to replicate it. The research conducted by Aronow during the 1970s had provided valuable independent confirmation of adverse health effects at low levels of exposure to CO. Aronow was the author of seven of the eight major studies relied on by EPA to support the 1980 proposal, and his work was clearly central to the development of a scientifically credible NAAQS for CO.

Warning signals concerning Aronow's work had appeared in the scientific literature before 1983. Other researchers had publicly questioned his studies on methodological grounds, such as inadequate controls and too few subjects, as well as because they were unreplicated.[66] EPA's staff was aware of these criticisms, but believed they were neutralized by the support Aronow received from other quarters, including a respected member of CASAC. As of 1983, EPA had no reason to doubt Aronow's integrity, although both FDA and the VA had concluded by 1982 that some of his clinical studies on new drugs contained "fudged" data. In fact, the VA had directed Aronow in 1980 to discontinue all research activities, and in 1982 FDA limited his right to serve as a clinical investigator for pharmaceutical companies.

These unwelcome discoveries led EPA to establish a peer-review committee to audit all of Aronow's studies cited in the CO criteria document and to recommend how far the agency could still rely on this work. Based on the committee's recommendations, EPA eventually eliminated all references to the Aronow studies except as a basis for determining the margin of safety for the standard. In the meantime, the agency sponsored new research to see whether Aronow's findings could be replicated[67] and sought additional scientific support for the proposed CO standard.

EPA's efforts to reestablish the scientific integrity of the CO standard received invaluable support from CASAC. Despite questions about some of the new studies identified as "key" in the criteria document, the committee concluded that there was an adequate scientific basis for maintaining the CO standard at approximately the same levels as in 1971. On several crucial issues, CASAC apparently agreed with EPA that doubts should be resolved so as to err on the side of safety. For example, the committee determined that one study that had generated some criticism in the scientific community did not have to be disregarded simply because it was unreplicated.[68] CASAC also endorsed EPA's judgment that studies demonstrating the effects of CO on healthy subjects could be factored into the standard-setting. In sum, the committee seemed prepared to accept the lowered standard of proof that EPA was forced to adopt after the Aronow debacle, thus giving a needed boost to the agency's credibility.

CASAC's Effectiveness: Bridging Science and Policy

Since the inception of EPA's air quality program, the agency's approach to seeking scientific advice on standard-setting has changed almost beyond recognition. The "before" and "after" studies of the ozone rulemaking, as well as the revision of the CO standard, demonstrate that CASAC, EPA's primary source of advice on air pollution control, has established a fundamentally sympathetic working relationship with agency staff. Criticisms are offered on the whole in a spirit of constructive guidance, based on a general acceptance of the agency's goals and mission, and committee meetings have become a forum for reconciling divergent interpretations of evidence. How did this fortunate state of affairs come about, and how is it likely to benefit EPA in the future?

The answer to the first question lies partly in the legislative developments that have restructured the scientific advisory process under the Clean Air Act. When the first ozone review was set in motion, Congress had not yet provided for periodic revision of criteria documents, accompanied by CASAC review. In the absence of set procedures for scientific review, EPA solicited advice first from SAB and then from the Shy Panel, hoping by these means to enlist some external scientific support for the 0.08 ppm standard. These actions, however, proved fatal to EPA's hopes of maintaining scientific credibility. By openly shopping around among experts, EPA exposed the "constructed" character of its rationale for action and simply opened the way for every other interested party—SAB, API, NRDC, the White House economists—to come forward with their own constructions of the evidence. As Melnick's analysis shows, it became virtually impossible to claim that EPA's reading of the science was entitled to any greater deference than the alternative readings supplied by other scientific and political actors.

The 1977 amendments to the Clean Air Act cemented CASAC's role as EPA's primary consultative body and essentially ruled out the possibility of doing an end run around the committee. At the same time, EPA came to accept the premise that review by CASAC should encompass issues that had previously been designated as science policy, and hence implicitly considered out-of-bounds for the advisory committee. EPA formally signaled its change of position on this matter by involving CASAC in reviewing not only criteria documents, whose scientific character was never seriously questioned, but also staff papers, which border much more closely on policy. This means, in practice, that the agency and its scientific advisers have ample opportunity, from the earliest stages of the standard-setting process, to negotiate their differences on matters straddling the fuzzy boundary between science and policy. This relatively conciliatory style of interaction reduces the possibility of unbridgeable intellectual rifts between the agency and its scientific advisers; at the same time, it allows EPA to enlist the authority of science (symbolically represented by CASAC) in support of possibly controversial decisions, such as how to characterize "at risk" populations. As we see in the following chapter, not all EPA programs have succeeded equally well in striking such a happy accommodation between their scientific and political imperatives.

7

Advisers as Adversaries

Scientific credibility has been harder for EPA to maintain in its pesticide program than in any other area of regulatory responsibility. The program's transfer to EPA was a legacy of Rachel Carson's *Silent Spring,*[1] a book that not only launched a new social movement but helped locate pesticides at the very heart of environmental politics. As the promise of bountiful harvests gave way to images of spring without birdsong, pesticides came to symbolize much of the late twentieth century's ambivalence about using technology to master nature. An infant EPA found its policymaking efforts battered from the start by conflicting ethical and political demands that were almost impossible to reconcile. If the agency curbed the use of pesticides it was perceived as caving in to naive romanticism and misplaced back-to-nature demagoguery. If it permitted continued use of suspected chemicals, it was seen as captive to powerful agribusiness interests. To project scientific self-confidence under these circumstances was difficult at best, and neither EPA nor its scientific advisers proved entirely equal to the task.

The state of knowledge about the risks and benefits of pesticides did not work to EPA's advantage. While anxiety over the use of pesticides continued to grow around the globe, scientific proof that they present identifiable, measurable risks to health or the environment remained surprisingly difficult to establish, especially at low levels of exposure. Two gaps in the evidence emerged as specially troublesome to regulators. First, there was a dearth of reliable epidemiological data on health effects. Decisions had to be made in most instances on the basis of animal studies, but the principles for extrapolating from animal experiments to humans remained endlessly controversial. Second, the precise levels of human exposure to pesticides proved notoriously difficult to establish even for workers who manufacture or apply the products, let alone for individuals exposed only through environmental pathways such as air, food, and water.

By contrast, the economic benefits of pesticides, measured in terms of reduced crop loss and increased yield, seemed more tangible, hence easier for policymakers to respect. EPA was not permitted to lose sight of these benefits under its governing statute, which requires a balancing of costs against benefits in determining whether pesticides will cause unreasonably adverse effects on the environment. Indeed, the framework of the Federal Insecticide, Fungicide, and Rodenticide Act (FIFRA) left considerable doubt whether Congress really wanted EPA to get tough on pesticides or whether it was more concerned about avoiding Type II errors, that is, decisions to regulate products that are, in reality, harmless.

Under these circumstances, EPA felt a definite need—perhaps even a compulsion—to assure industry and the public that its decisions were scientifically informed and balanced and that the extent of risk was not overstated. The establishment of the Scientific Advisory Panel (SAP) and later expansions of its supervisory duties seemed to give the agency an institutional mechanism for maintaining a credible scientific posture. Yet an evaluation of three major regulatory controversies, involving ethylene dibromide (EDB), dicofol, and Alar, shows that this mechanism was far from perfect. Instead of facilitating scientific consensus-building, the SAP advisory process created a fundamentally adversarial environment, in which conflicting constructions of science were pitted against each other. In the process, the panel dealt a blow not only to EPA's scientific authority, but, ironically, to its own.

The Scientific Advisory Panel

In providing legislative authorization for the SAP, Congress gave back to the agriculture lobby a little of the ground it lost when pesticide regulation was removed from the U.S. Department of Agriculture (USDA) and placed within the newly established EPA. Agricultural interests, fearful that EPA was adopting an overly risk-averse interpretation of its legal mandate, vociferously argued for independent scientific review of EPA's decisions under the 1972 FIFRA amendments.[2] Congress responded positively, tacitly acknowledging that the complex task of sorting out the factual and methodological arguments about pesticide hazards could not be left exclusively to EPA's administrative discretion.[3] Environmental groups, who had unanimously objected to the SAP, failed to persuade the legislature that FIFRA already provided ample opportunities for scientific comment and that mandatory referral to outside experts would only worsen an already backlogged administrative system.[4]

Initially, FIFRA required consultation with SAP for cancellation decisions under Sec. 6(b) and for regulations promulgated under Sec. 25(a).[5] In 1978 Congress stipulated more generally that EPA should solicit advice on guidelines for improving the effectiveness and quality of scientific analyses leading to regulatory decisions.[6] Other responsibilities were added in 1980 in response to EPA's controversial decision to bypass the advisory panel in suspending 2,4,5-T.[7] Specifically, the 1980 amendments provided that the administrator should seek the advisory panel's comments on any suspension decision to prevent an imminent hazard.[8] Second, the administrator was required to develop written procedures for peer review by the SAP of "major scientific studies" carried out by or for EPA under FIFRA.[9]

FIFRA provides a relatively formal process for selecting SAP members. Six candidates each are nominated by the National Institutes of Health and the National Science Foundation, and the EPA administrator selects seven panelists from the resulting list.[10] Vacancies are filled through a similar procedure, with NIH and NSF each submitting two names, and the administrator selecting a replacement from the group of four nominees. In theory, the members' terms are staggered, so that all appointments to the panel do not expire at once. The 1983 amendments provided, further, that panelists should be selected to include expertise in toxicology, pathology, environmental biology, and other sciences relevant to assessing the health or environmental risks of pesticides.[11] Maintaining an appropriate disciplinary balance on the panel, however, proved to be a difficult task. In its early years, the SAP was hampered by a shortage of toxicological expertise;[12] later, it suffered from weaknesses in the statistical aspects of risk assessment, in ecotoxicology, and to a lesser degree in the area of wildlife effects.

While Congress was happy to expand SAP's powers, it was noticeably more reluctant to secure continuity in SAP's relationship with EPA. Thus, in 1978 a House-Senate conference committee elected to extend the panel only to September 30, 1981, instead of making it a permanent part of EPA, as recommended by the House.[13] Unforeseen delays in reauthorizing FIFRA caused SAP's legislative mandate to expire in 1981. Faced with the panel's demise, EPA reestablished it with identical responsibilities under FACA.[14] Although appointment letters were sent out in 1982, the reconstituted SAP did not meet again until June 1983, at which time the panel was composed entirely of new members.[15] This disruption drastically altered the panel's work pattern. Prior to 1981, SAP met regularly once a month except when agency business was too light to justify a meeting. After SAP reconvened in 1983, however, it met only twice in the first year and the frequency of meetings did not pick up substantially after that date.

SAP's style of operating in the years after its formation caused concern that the panel might evolve into a hearing board rather than a body limited to reviewing the scientific basis of EPA decisions. The panel's meetings were open to the public, and SAP occasionally allowed these sessions to blossom into adversarial hearings on policy issues ranging far beyond the review of agency risk assessments.[16] Given SAP's broad construction of its powers and its willingness to hear from interested parties, some feared that the panel would be captured by industry through aggressively mounted presentations by expert witnesses. But despite such concerns, SAP's relations with EPA remained cordial and no major rifts developed over the interpretation of scientific data or its application to regulatory proposals. According to one estimate, SAP in this period approved about 90 percent of the assessments performed by the staff of EPA's Office of Pesticide Programs (OPP).[17]

On at least one occasion, moreover, industry's attempt to win over the SAP through an immoderate attack on EPA's science produced exactly the opposite effect. The American Wood Preservers' Institute challenged EPA's decision to restrict the permitted uses of the preservatives pentachlorophenol, inorganic arsenicals, and creosote at a SAP meeting in June 1981. The panelists were struck by "the lack of scientific objectivity of the presentations by industry" and found "the industry's denial of scientific data concerning the mutagenicity and carcinogenicity of the wood preservatives to be disturbing."[18] After weighing the Institute's arguments against EPA's, SAP agreed with the agency that creosote is a mutagen, but that more epidemiological studies would be needed to establish a carcinogenic risk to humans. But the period of cooperation was short-lived. Over the next few years, SAP and OPP clashed sharply on several occasions, and their disagreements coincided with a rise in public uncertainty and anxiety about pesticide safety.

Implementing the Impossible

When Congress amended FIFRA in 1972, it set EPA some daunting tasks. The new law not only changed the legal standard for registering new pesticides[19] but also required EPA to reregister some 50,000 products that were already on the U.S. market and to cancel the registration of any pesticide whose effects were deemed unreasonably adverse. The four-year deadline originally provided for this task soon proved illusory, and it was eventually repealed. But even without the pressure of statutory deadlines, OPP had to devote a considerable portion of its energies to devising a rational strategy for reregistering "old" pesticides.

Progress was hindered at the outset by an overreliance on formal adjudicatory hearings. EPA began by sending out notices of intent to cancel the registration of any pesticide that did not appear to meet the statutory safety standard.[20] This was followed by trial-type proceedings to develop evidence concerning the product's risks and benefits. These proceedings proved not only costly and cumbersome but exclusionary, since only the parties who were formally represented could participate in presenting and critiquing evidence. An internal agency task force joined critics outside the agency[21] in calling for a more informal process that would expedite review and provide better opportunities for public participation. In 1975 Congress introduced other actors into the review process through amendments to FIFRA. The law now required that proposed cancellations or changes in classification be submitted to USDA[22] and to the newly created SAP[23] before notices of intent were sent to registrants. These procedures were designed to ensure that EPA would look closely at both risks and benefits prior to initiating cancellation hearings.

EPA, meanwhile, designed a modified review process called the Rebuttable Presumption Against Registration (RPAR) in order to streamline the analysis of risks and benefits. Regulations defining how products would enter RPAR review and how they would be further evaluated were published in 1975. As a threshold matter, each pesticide was measured against a series of "risk criteria" governing acute and chronic toxicity, hazards to wildlife, and the availability of emergency treatments.[24] These criteria screened out pesticides posing a substantial question of safety (and hence suitable for further RPAR review) from those that could be considered reasonably safe. The RPAR process continued with an informal comment period during which interested parties could either rebut the presumption of unreasonable risk or seek to establish the hazard more firmly. EPA initially expected this comment period to be brief and to lead relatively quickly to a final regulatory decision.[25] In practice, however, the RPAR process proved scarcely less time-consuming than the prior regulatory approach, though it may have lowered EPA's personnel and legal costs. Three to seven years were ordinarily required for the completion of a full RPAR review.[26]

FIFRA requires that existing products may be taken off the market only on the basis of validated tests or other significant evidence indicating an unreasonable risk.[27] Accordingly, during the initial investigation phase of the RPAR process, EPA had to review and validate all scientific studies suggesting that the product under review might meet or exceed one or more of the risk criteria. The validated studies, together with the results of a literature search and information on uses and routes of exposure, formed the basis for Position Document 1 (PD 1), a summary of the agency's preliminary assessment of the risk of the pesticide. Subsequently, the registrant had an opportunity to rebut the presumption of risk by challenging the scientific validity of the agency's studies or by reducing the risk to levels that were not "unreasonable." If EPA was persuaded that the pesticide was reasonably safe (that is, did not meet or exceed the RPAR criteria), it was returned to the normal registration process; this action was described in Position Document 2 (PD 2), which ended RPAR review for the product in question. If the rebuttal was unsuccessful, however, the agency prepared Position Document 2/3 (PD 2/3), which incorporated EPA's analysis of risks and benefits, evaluation of regulatory options, and proposed regulatory position.

Pursuant to the 1975 FIFRA amendments, PD 2/3 had to be reviewed by the SAP and the Secretary of Agriculture. In addition, a notice of the availability of this document was published for public comment. The agency considered responses from SAP, USDA, the industry, and other interested parties in reaching its final decision. A fourth position paper (PD 4) contained the risk-benefit evaluation that supported the proposed action.

Beginning in 1985, EPA began considering the first major revisions to the RPAR program in a decade.[28] Industry had for some years been dissatisfied with the term "RPAR." Registrants complained that a "rebuttable presumption against registration" implied—misleadingly, in their view—that a product was unreasonably dangerous, in other words, that its risks were not justified by its benefits. Yet EPA's threshold determination that a rebuttable presumption exists was made before the agency had conducted any review of the benefits. Manufacturers were fearful that the term RPAR might send the wrong signal to foreign governments or members of the public who were unfamiliar with EPA's full decisionmaking process. To set these misgivings to rest, EPA agreed to change the name of the RPAR process to the more neutral "Special Review."

The special review regulations made no fundamental changes in the procedures for soliciting scientific advice.[29] Under the new rules, a formal Notice of Special Review is issued for pesticides that meet or exceed the risk criteria. This is followed by a comment period corresponding to that which followed the preparation of the PD 1 in the RPAR process. If, at the close of comments, the agency determines that it is necessary to cancel or deny a registration, to hold a hearing, or to change a classification, a preliminary determination with respect to these issues must be submitted to the SAP and USDA for comment. The Notice of Preliminary Determination corresponds to PD 2/3 in the RPAR process. As before, SAP's responses are incorporated along with other public comments into the agency's final determination concerning the pesticide.

EPA developed another mechanism to streamline the reregistration of old pesticides: the Registration Standards Program. Recognizing the near impossibility of reregistering 50,000 separately registered compounds, Congress in 1978 authorized the agency to use a generic-standards approach in dealing with these substances. Instead of addressing each registered formulation individually, EPA could now group them according to the active ingredients that they contain. Approximately 600 registration standards, each representing a single active ingredient, suffice to cover all currently registered pesticides. In order to speed the review of these ingredients still further, EPA grouped them into 48 "clusters." The clusters were ranked according to a priority scale that took account of production levels, human exposure, and ecological exposure.

Though the generic approach to standard-setting reduced the number of reviews from a mind-numbing 50,000 to only 48 clusters of proceedings, even this more bounded task strained EPA's regulatory capabilities. As of June 1984 EPA had completed work on 76 of the 600 standards and was proceeding at the rate of 25 new standards per year.[30] A hostile congressional committee noted that, at this rate, the agency would not finish issuing the standards until the year 2005. Moreover, setting a generic standard is not equivalent to completing the registration of all products containing the corresponding active ingredient. The standards program identifies important data gaps for individual products. To complete the registration, registrants must still supply missing data, as well as comply with composition, labeling, and packaging requirements, as specified by the registration standard. By April 1983 EPA had reregistered only 70 pesticides representing 4 active ingredients. Work on another 360 products was delayed pending litigation over FIFRA's registration and data release provisions. A House investigative report observed that even if EPA succeeded in reregistering all of these compounds, it would have tackled less than one percent of the products subject to reregistration.[31] With its credibility under constant attack in Congress, and constrained by multiple forms of oversight, OPP confronted the regulatory science controversies of the 1980s from a position of singular weakness.

Ethylene Dibromide

The ethylene dibromide (EDB) crisis represented the first major test of OPP's ability to work with SAP in the deregulatory environment of the Reagan administration. The case is noteworthy both because it sowed the seeds of later, more consequential conflicts between the agency and its scientific advisers and as an object lesson in the limitations of regulatory peer review. Timed to occur after political considerations had already become paramount in the decisionmaking process, scientific review proved at best irrelevant and, at worst, an infringement upon EPA's policy authority.

The Birth of a Public Problem

EDB captured the nation's newspaper headlines in the winter of 1983–84, but the story began in fact many years earlier. A halogenated hydrocarbon, EDB had been used as an insecticide since 1925 and had been registered as a pesticide by USDA in 1948.[32] As of 1984, more than 300 million pounds of the compound were produced annually in the United States.[33] By far the largest share (about 230 million pounds) went into leaded gasoline as an additive to clean lead deposits out of motor vehicle engines. Of the remainder, some 20 million pounds were used in agriculture, primarily as a soil fumigant for nematode control. The pesticide was directly injected into the ground, and soil so treated was used to grow more than thirty fruit and vegetable crops in California, Hawaii, and the southern states. There was a big increase in usage in 1980 when EPA granted emergency approval for soil fumigation with EDB prior to planting soybeans.[34] EDB replaced the cancelled pesticide dibromochloropropane (DBCP) which had previously been used for this purpose. The most important remaining uses of EDB were as a fumigant for grain, grain milling machinery, citrus, papaya, and other tropical fruits.

EDB's toxicity had been recognized since 1927, when reports of adverse effects in exposed animals first began appearing in the literature.[35] By 1973 a number of studies had reported that EDB caused mutations and reproductive damage in animals. In 1974 NCI issued a "memorandum of alert" warning that its ongoing studies had identified EDB as a potential carcinogen. A final report published in 1975 concluded that EDB should be considered capable of causing cancer in humans. David Rall, director of the National Toxicology Program, told a reporter in 1984 that EDB had demonstrated carcinogenic effects at the lowest tested levels in three major government studies, producing cancer in virtually 100 percent of the test animals.[36] Though EDB was originally believed not to accumulate in the environment or the food chain because of its volatility, improvements in measurement techniques revealed traces in food by the late 1970s, with levels in some wheat samples reaching 4200 ppb (parts per billion).[37] FDA began a program to monitor EDB residues in food in 1978.[38] Between 1982 and 1983, EDB was also detected in groundwater in several states.[39]

The findings of carcinogenicity and other health effects, as well as the discovery of EDB traces in food, led EPA to issue an RPAR notice for the substance in December 1977.[40] Thereafter, the regulatory process moved at a glacial pace, producing no visible results for three full years. EPA's inability to reach agreement with USDA and with the chemical manufacturers on what to do about specific uses of EDB probably accounts for this lack of progress. In December 1980, just before the Carter administration disbanded, EPA prepared a proposal to cancel EDB's uses as a fumigant for grain and grain milling machinery and to phase it out as a citrus fumigant over the next two years.[41] While EPA also expressed reservations about EDB's use in soil fumigation, no action was recommended at this stage, since, before the discovery of groundwater contamination, this use was not believed to result in significant human exposure. By April 1981 the SAP had reviewed these proposals and had submitted a final report concurring with the cancellation decision. However, SAP considered the phase-out of citrus fumigation to be unnecessary, as well as impracticable, because there were no substitutes available for this purpose; the panel suggested that the risks of that use could more appropriately be handled through increased protection for exposed workers.[42] It appeared that there were no further scientific obstacles to EPA's developing a final regulatory proposal for the pesticide.

Yet for the next two years EPA took no action on EDB. In 1982 the proposed regulatory package was transmitted to the office of John Todhunter, then director of OPP and a Reagan appointee, where it "inexplicably disappeared"[43] until Todhunter's departure from the agency. Moreover, Todhunter unaccountably requested the SAP to review the EDB regulatory package again in 1983. The stalemate on EDB was part of a larger picture of inaction at OPP in the initial years of the Reagan administration. Todhunter slashed the staff for carrying out special pesticide reviews from 128 to about 20.[44] No new RPARs were issued between April 1981 and March 1984, although OPP continued to process RPAR actions commenced in the preceding four years.[45]

In the meantime, the political picture with respect to the compound was changing rapidly. In 1981 the state of California embarked on a massive EDB fumigation program to control and quarantine the Mediterranean fruit fly (Medfly).[46] The state Occupational Safety and Health Administration (Cal OSHA) countered by proposing a stringent new workplace standard for EDB and by requiring warning placards to be placed in areas contaminated with EDB fumes.[47] The resulting controversy over worker safety soon acquired both interstate and international overtones. Japanese dockworkers refused to unload fruit arriving in their country until the Cal OSHA exposure standard was adopted. The Japanese government, however, was more concerned with keeping out the Medfly and insisted on fumigation of all produce arriving from California. California supermarkets, in the meantime, began a boycott of EDB-fumigated fruit from Texas and Florida so as to avoid any difficulties with produce that did not meet Cal OSHA requirements. State activism reached another peak in late 1983 with Florida's decision to ban the sale of more than seventy grain products containing residues of EDB.[48] By this time, the federal government was again at work on EDB.

In the summer of 1983 EPA was feverishly developing an up-to-date regulatory package on the pesticide, spurred largely by the discovery of EDB in groundwater. Several agency actions followed in quick succession, eventually ending almost all agricultural uses of the compound. In September 1983 EPA suspended EDB as a soil fumigant.[49] In February 1984 the agency announced an emergency suspension of EDB as a fumigant for grain and grain milling machinery, and set residue limits on raw and processed grain.[50] According to the new EPA administrator, William Ruckelshaus, these actions were designed to clear EDB from the nation's "food pipeline."[51] The agency's final action was to establish interim tolerances for citrus and papaya, and to order that all residues on these fruits should be eliminated by September of 1984.[52]

Scientists as Policy Analysts

While these actions helped allay public concern,[53] EPA still had to win the support of its scientific advisers, and this proved far from straightforward. When EPA presented the matter to the SAP in November 1983, one issue that particularly troubled the panel was the state of knowledge about the alternatives to EDB. Members repeatedly questioned OPP representatives about the toxicity and the environmental transport and fate of other likely soil fumigants.[54] There was frank skepticism that Telone, the only viable nematocide left after the cancellation of DBCP and EDB,[55] would prove any safer in the long run. As one panelist observed, "I think it's very important in the whole deliberations of this EDB problem that we don't just get rid of this stuff and jump into assuming that Telone is going to take over, then we are going to be back here two years down the road."[56]

In June 1984 the panel met again to review the emergency suspension of the grain fumigation uses of EDB. Again, one of SAP's principal concerns was that the agency had paid insufficient attention to the hazards of probable alternatives, particularly methyl bromide. OPP staffers conceded that as of the PD 4 on EDB the toxicology of methyl bromide was not well known, but that a Dutch study had since concluded that this compound, too, was an animal carcinogen.[57] In the agency's view, however, methyl bromide, being more volatile than

EDB, could be expected to pose fewer risks of environmental contamination and exposure. This reasoning did not reassure the panel, one of whose members commented:

> The fact that it is more volatile might well create a high degree of hazard in those who apply it and are exposed to it that way even though it doesn't get into the food chain and the finished products as much. It concerns me that we may simply just be creating another scare a year from now, or a couple of years down the road by going to these materials.[58]

Another member echoed these remarks, saying, "You can see the whole EDB thing being repeated again, and some of these food uses and grain milling uses, the same sort of thing again."[59]

These exchanges show the SAP focusing on an area of agency policy that fell outside the panel's official jurisdiction: EPA's approach to assessing the risks of compounds related to EDB, but not yet in use as alternatives. While SAP members agreed that EDB was a potent carcinogen and that its removal from the food supply was prudent,[60] they balked at the prospect of considering EDB and its alternatives case by case, possibly setting in motion the "short line of dominos" of which one critic warned.[61] A panel member summed up his colleagues' reservations:

> A lot of us feel this is a generic problem related to nematocides. To have good nematocides, you have to have compounds with this general type of property . . . It's a generic problem, and I think the Agency ought to think of it as a generic problem, as opposed to thinking of a single chemical like EDB.[62]

The panelists also felt that EPA's efforts to obtain exposure data on EDB and both exposure and effects data on alternatives had been unreasonably slow. Why, inquired one member, had it taken ten years from the appearance of preliminary bioassay results on EDB for the agency to start giving systematic consideration to its alternatives?[63] Another panelist was reluctant to blame EPA for the deficiencies in the data, but noted that it should have been possible for any competent chemist to look down the line from DBCP to EDB and Telone and to make sure that the data would be on hand to assess them.[64]

Complaints about the slow pace of EPA's data collection efforts and regulatory movements were coupled with annoyance over what some panelists saw as unnecessary hysteria over EDB:

> You know . . . it is hard to conceive that an emergency suddenly existed . . . it couldn't have waited at least for our group to meet

> with you people to give our opinion. As far as I know, no consumer has died from EDB. Nobody has gotten cancer. Now you are substituting chemicals which offer the possibility of severe toxicity, if not death of the applicators.[65]

In extenuation, OPP representatives pointed out, first, that an "emergency suspension" did not reflect a desire to create a mood of emergency, but rather that such action was authorized by FIFRA where cancellation proceedings threatened to consume too much time.[66] Agency staff also noted that they had not been authorized to "call in" missing information with respect to existing registrations until Congress amended FIFRA in 1978.[67] However, one SAP member rejoined that this meant the agency had had the relevant authority for six years before undertaking to use it on substitutes for EDB.

The Risk-Assessment Controversy

A more technical divergence developed between OPP and the panel with respect to the issue of quantitative risk assessment. FIFRA's requirement of balancing risks and benefits invites EPA to quantify health risks so as to make them more commensurable with economic costs to industry. As we have already seen, however, carcinogenic risk assessment involves multiple conflicts and uncertainties.[68] In the EDB case, EPA attempted to overcome these uncertainties with a numerically precise risk estimate, but this effort backfired when the agency was forced to explain itself to skeptics on the advisory panel.

Estimates of the cancer risk from exposure to EDB have varied by several orders of magnitude depending on the risk-assessment model and the assumptions used. The approach used by EPA in PD 4, for example, predicted an additional risk of 3.3×10^{-3} (or one cancer per 300 population) from the consumption of fumigated grain.[69] The numbers looked incomparably more benign in a risk assessment commissioned by the Grocery Manufacturers of America from an independent Washington consultant, Joseph V. Rodricks of Environ Corporation. His analysis yielded a risk to adults of about one in 12 million and to children of one in 4 million.[70] The principal reason for this divergence was that EPA estimated average residue levels of 31 ppb in grain, whereas Rodricks, using allegedly more complete data provided by his client, used a value of only 0.033 ppb. In the meantime, the states of New York and California found that their own risk assessments produced numbers of even greater concern than those generated by EPA.[71] Their conclusions, in turn, influenced other states that had much less expertise in the area of carcinogenic risk assessment, such as Massachusetts. For example, Dr. Bailus Walker, Massachusetts Commissioner of Public Health, based his state's policy with respect to EDB on a New York assessment that put the risk of cancer to children at 1,600 per million from the proposed EPA standard of 30 ppb for grain products; the same assessment showed a risk of only 50 per million from a 1 ppb standard.[72]

The inexactitudes of risk assessment troubled members of the SAP, many of whom believed that the decisions concerning EDB should simply have been based on a qualitative evaluation of the compound's effects on health. The panel's position was perhaps most forcefully articulated by Edward Smuckler, a veteran member of the SAP:

> The point I would like to call your attention to is the fact that this material is indeed a potent oncogenic. It affects a number of systems, regardless of routes. It affects a number of species, regardless of routes and also reproductive effects. It certainly is a very biologically active agent. It seems to me worrying about the gimcrack mathematics used to estimate human risk, which I believe are fraudulent, too many assumptions go into it. This material is indeed a biologically potent active agent and should be addressed as such without worrying about quantification that does not have any substance in fact.[73]

Two other panelists suggested that the toxicological data were in any case inadequate to support the kind of mathematical manipulation required for risk assessment. Christopher Wilkinson, asserting that he believed EDB to be carcinogenic, asked, "Why does the Agency continue to perpetuate . . . this myth that you can describe in some way these very marginal data . . . by these statistical models and risk analyses that, frankly, don't mean a thing?"[74] (Wilkinson did not explain why the "marginal data" supported the qualitative conclusion, shared by himself and Smuckler, that EDB was a potent carcinogen.)

An exchange between Wendell Kilgore and OPP staff questioned not merely the adequacy but the relevance of the toxicological data base in even more graphic terms:

Dr. Kilgore: I am wondering if you can tell me what type of cancer EDB causes in man.
Mr. Cohen: No, I can't.
Dr. Kilgore: Can anyone from EPA? How can you calculate what

incidence would be in man if you don't know what kind of cancer it causes in man?

Dr. Johnson: The risk estimates that were developed and Todd Thorslund put together there were for the specific cancer sites.

Dr. Kilgore: Cancer in the forestomach of rodents?

Dr. Johnson: That's right.

Dr. Kilgore: But man doesn't have a forestomach, I don't think.

Dr. Johnson: That's right. They have an esophagus though.[75]

Kilgore's comments here make sense only if one believes that chemically induced cancer is always site-specific (that is, occurs at the same site for every species). This view, however, was flatly at odds with EPA's science policy determinations concerning this issue.[76] The agency has long held that an animal carcinogen should be treated as a human carcinogen for purposes of risk assessment, regardless of the site at which cancer occurs in test animals. Kilgore's remarks also suggested a general reluctance to use animal data in the absence of supporting evidence about humans. This, again, was at odds with EPA policy.

Chairman Menzer's summary of SAP's position on the following day also harped on the theme of inadequacies in the data. The panel, Menzer noted, saw the agency's quantitative risk analysis as "an attempt to use statistical models to convert inadequate toxicological data into precise values for human risk."[77] In the case of EDB, the panel did not believe it was necessary to make such numerical predictions in order to support EPA's policy decisions. As Menzer further observed, "The fact that EDB is a potent animal carcinogen is enough to make its exposure in man undesirable." But the panel's judgments implied that this particular group of scientists would look with far greater skepticism on regulatory actions based on weaker toxicological and epidemiological data. This, indeed, was exactly what happened in the cases of dicofol and Alar.

Dicofol

As in the EDB case, EPA's decisionmaking on dicofol (also known as Kelthane) was shaped by challenges from two different directions: pressure from environmentalists in and out of Congress to move faster on recognized hazards, and criticism from the industry and the SAP for inadequate attention to problems in the scientific record. The product in this case was a miticide manufactured in Israel and distributed by Rohm and Haas in the United States, where some 2 to 3 million pounds of the compound were used each year on a variety of crops, particularly cotton and citrus fruits.[78] Dicofol attracted attention in March 1984 as the first new chemical to be selected for RPAR (or special review) following the hiatus on such actions between 1981 and 1984.[79] EPA's primary concern was that products containing dicofol also contained DDT and structurally similar compounds as impurities. DDT was banned in the United States in 1972 because its accumulation in the environment posed hazards to wildlife and because the substance was identified as a potential human carcinogen.

Dicofol attracted political notice when a subcommittee of the House Committee on Government Operations heard testimony about EPA's handling of the compound at the pesticide registration hearings of 1983 and 1984.[80] This was a forum in which environmental interests generally predominate over agricultural interests. Committee members professed great concern about what they saw as an unconscionable delay between EPA's banning of DDT and the discovery that dicofol preparations were contaminated with related compounds. Dicofol manufacturers had submitted confidential statements to the government since 1957 showing DDT as an impurity. Yet this fact was not officially noticed by EPA until the agency initiated a Registration Standard review of dicofol in 1979.[81] According to EPA's own regulations, the presence of DDT, a banned product, should have been sufficient by itself to trigger RPAR review; however, the agency in fact waited until it had completed the registration standard process for dicofol. The congressional investigators seemed unimpressed by EPA's argument that completing the registration standard first led to an administratively more efficient decision. Under pressure to take concrete action on dicofol, EPA in late 1984 published a notice of intent to cancel the pesticide's registration.[82]

An additional cause for concern at EPA was a 1978 NCI bioassay showing positive carcinogenic effects in male mice, though negative results were obtained in female mice and in rats of both sexes. By January 1985 EPA's Carcinogen Assessment Group (CAG) had completed an analysis of the carcinogenic hazards of dicofol. The CAG report noted that the equivocal findings of the NCI study would ordinarily produce a relatively low risk classification for dicofol. However, because of the compound's structural similarity with DDT, a substance for which CAG claimed to have much stronger evidence of carcinogenicity, dicofol was classified in a higher risk group.[83] CAG also carried out a quantitative risk assessment based on the cancer response in male mice. The resulting estimate of the human cancer potency of dicofol was roughly comparable to that of compounds in the DDT family.[84] On the basis of these findings, CAG recommended that technical grades of dicofol be considered potential human carcinogens for risk-management purposes.

CAG's conclusions were vigorously contested by Rohm and Haas at a SAP meeting in February 1985. The crux of the company's case was that there was no scientific support for classifying dicofol as a carcinogen. To get this viewpoint across to the SAP, the company marshaled its experts with considerable care and skill. The core of Rohm and Haas's presentation was an "independent" peer review of the carcinogenicity data, prepared by three experts designated by the company as its "pathology panel." One member of this panel was Edward Smuckler, who had participated in reviewing EDB as a SAP member only a few months earlier. The two remaining members also had impressive credentials in pathology. All three panelists made personal evaluations of the slides from the NCI bioassay, and on the basis of these examinations, they unanimously concluded that dicofol did not produce cancer in mice, contrary to the conclusions of the NCI pathologists. Applying allegedly more up-to-date diagnostic criteria than those employed by NCI,[85] the panel asserted that the effects observed in the male mice were "non-neoplastic nodules" which did not progress into either benign or malignant tumors. The industry experts also took issue with CAG's determination that dicofol's genotoxic potential could not be evaluated from the available data.[86] They found rather that there was no evidence of genotoxic activity in any of the large number of mutagenicity studies conducted on the compound.

With respect to EPA's quantitative risk assessment, Rohm and Haas claimed that, in view of its pathology panel's qualitative findings as to carcinogenicity, CAG should never have carried out such an assessment for dicofol. Company experts further argued that a properly performed quantitative assessment would have exonerated their product altogether. Frank W. Carlborg, a private statistical consultant and expert on risk assessment, estimated the risk of cancer from dietary exposure to dicofol to be one in 10 million, or well below levels of concern, even if one used worst-case assumptions.[87] Similarly, the risk to applicators was estimated to be one in a million, a level that EPA has often treated as insignificant.[88] In making these calculations, Carlborg took issue with several of CAG's key assumptions, including the use of surface area rather than body weight for extrapolating the carcinogenic response from mice to humans, and the use of the linearized multistage model rather than the Weibull model for low-dose extrapolation.[89]

Leaving nothing to chance, Rohm and Haas also followed the classic legal strategy of arguing in the alternative. Even if the NCI study provided valid evidence of carcinogenicity, the firm suggested, dicofol still should not have been classified as a carcinogen in accordance with EPA's proposed guidelines for risk assessment.[90] The company also questioned CAG's classification of the related compound DDT as a probable human carcinogen. Finally, Rohm and Haas announced a plan to reduce the levels of DDT-like compounds in its dicofol preparations from as much as 10 percent to less than 0.1 percent within two years.[91]

The upshot of this carefully mounted presentation was precisely what the company had intended. SAP was persuaded that dicofol should not be viewed as a human cancer hazard, and EPA decided to permit the use of dicofol with more than 0.1 percent DDT and related impurities for two more years as requested by Rohm and Haas. The decision seemed at one level to accomplish just what Congress wanted; timely peer review arguably kept overzealous agency scientists from adopting an alarmist risk analysis that was not supported by reliable data. Yet the dicofol proceedings were a far cry from the open, balanced, and dispassionate review process that Congress ostensibly sought to establish under FIFRA. The tactics used by Rohm and Haas showed how easily the authority of peer review could be coopted by an interested party wishing to advance its own construction of the scientific evidence. Given the impressive credentials of the company experts, and the doubly impressive consensus within the pathology panel, it was almost a foregone conclusion that the SAP would give greater credence to Rohm and Haas's position than to the arguments developed by relatively anonymous scientists in CAG, OPP, and NCI.

The result, however, was that the possible biases inherent in such "captive" peer review were never openly explored. Although the Rohm and Haas presentation was adversarial in style—the company's written docket submission was organized and argued like a legal brief—there was no equally systematic rebuttal. If the proceedings before the SAP were structured as a "battle of experts," it was a battle in which only one side fielded an army. As a result, there was no public testing of the Rohm and Haas experts with respect to the content of their submissions or their overall credibility. For example, no one questioned the propriety of the company's presenting as an "independent" expert at least one scientist (Smuckler) who may have had artificially high credibility with his former colleagues.

Alar

For several weeks in the spring of 1989, the nation was caught up in an unusual frenzy over the safety of apples treated with Alar (trade name for daminozide), a plant growth regulator registered since the 1960s for use on apples, peanuts, and other fruit. EPA and NRDC clashed publicly over the risk of cancer from Alar residues in treated fruit. Oscar-winning actress Meryl Streep appeared before a Senate Labor and Human Resources subcommittee with a plea to end experimentation on children,[92] while schools and supermarkets hastened to stop sales of commercially grown apples. A group of sixty-five scientists countered with a full page advertisement in the *New York Times,* decrying the "unfounded attacks on the safety of our food supply."[93] Finally, in June, Uniroyal, Alar's manufacturer, announced that it would no longer market the product in the United States. These extraordinary events represented at one level a complete breakdown in OPP's efforts to project scientific authority; at another level, they provided an instructive study in the social construction of science.

A Question of Risk

For the apple industry, which accounts for about 75 percent of Alar's total use, daminozide was close to a miracle product. It delayed ripening and extended harvest time, reduced preharvest drop, increased the shelf life of harvested fruit, and yielded fruits that were redder and firmer than untreated crops. But to OPP the chemical presented another, less benign aspect. Studies indicated that both daminozide and its breakdown product unsymmetrical dimethylhydrazine (UDMH) caused cancer in animals, and residues of both chemicals were found in treated fruit, particularly in cooked apple products, such as apple sauce.

Evidence that UDMH is an animal carcinogen appeared in the

scientific literature in 1973, when Bela Toth and colleagues at the Eppley Institute of the University of Nebraska Medical Center published a study of the chemical's effects in mice in the *Journal of the National Cancer Institute.*[94] Subsequently, at least two additional studies by the Toth group, a study by NCI, and a study done for the Air Force all indicated that UDMH and daminozide produce tumors in both rats and mice by different routes of exposure.[95] The Toth studies led an EPA scientist to propose in 1977 that daminozide be selected for special review. EPA notified Uniroyal, the principal manufacturer of the compound, that special review might occur, but did not actually initiate review. During the next three years, the agency issued data call-in notices to two daminozide registrants, Uniroyal and Aceto Chemical Company, requesting additional information on the conversion of daminozide to UDMH and on residue levels in agricultural products. Aceto did not comply, and its registrations were suspended under FIFRA Section 3(c)(2)(B). Uniroyal voluntarily canceled many of the minor uses of daminozide, thus avoiding some of the data requirements imposed by EPA.[96]

By mid-1984 EPA had completed the Registration Standard for daminozide, delineating gaps in the data on residues, chronic toxicity, and mutagenicity, and in July the agency announced that it would begin special review based on the possible carcinogenicity of both the parent compound and UDMH. Further, in 1985 EPA completed an audit of the Toth study of daminozide published in *Cancer Research* in 1977 and confirmed that there were significant increases in blood-vessel cancers in the tested species.

OPP's normal course at this stage would have been to issue a PD 2/3 describing its proposed regulatory action with respect to daminozide. OPP's plans, however, were partially derailed by concurrent developments at the Office of Policy, Planning and Evaluation (OPPE), the EPA unit charged with estimating the economic benefits of pesticides and balancing their costs and benefits. Having overcome numerous analytical difficulties,[97] OPPE staff had developed a model that they believed adequately captured the considerable costs of restricting daminozide's use. By OPPE's estimate, growers could expect to lose $32.9 million by abandoning daminozide treatment; the total cost to society, including losses to consumers, was estimated at about $60 million.[98] Nevertheless, OPPE economists came out in favor of expeditious regulation, for the risks of daminozide appeared to outweigh the costs by a significant margin. The pesticide office's risk-assessment data indicated that daminozide was as risky as EDB, which the agency had just decided to suspend, or perhaps even more so. Banning daminozide therefore seemed to be one of the cheapest regulatory options the pesticide program could pursue in terms of dollars per life saved.[99]

Opinion within the agency split between those who favored an immediate emergency suspension of daminozide, based on the carcinogenicity data and the preliminary risk-benefit calculations, and those who preferred to act more cautiously.[100] The two sides compromised on a decision to seek an "expedited cancellation," an ad hoc procedure that the agency hoped would stop daminozide applications before the 1986 growing season. An unusual composite position document, PD 2/3/4, was prepared to explain this decision and was submitted to the SAP for review at a meeting in September 1985.[101]

The SAP Proceedings

As in the dicofol case, the format of the presentation before the SAP was fundamentally adversarial, pitting seven OPP and Office of Toxic Substances (OTS) staff members against a panel of eight Uniroyal representatives. These included two company officials (the director of research and development, and the manager for registration and toxicology) and six consultants, including two who had appeared on behalf of Rohm and Haas in the dicofol proceedings (Stan D. Vesselinovitch, Frank W. Carlborg). Uniroyal's Washington counsel, Kenneth Weinstein, coordinated and managed the expert presentations for the company. Although the SAP meeting was open to the general public, the five additional speakers who attended the Alar session constituted a solidly pro-industry bloc: two growers, a Uniroyal toxicologist, the director of the National Food Processors Association, and a retired academic who had worked for apple growers in Virginia. There was no input from environmental or public interest groups who might have been expected to support EPA's proposal as a precautionary "prohealth" initiative.

As EPA and Uniroyal rehearsed their scientific arguments for the panel's benefit, it became clear that the two sides viewed the same data through entirely different analytical lenses. For EPA, the point of overriding importance was the coherence of the carcinogenicity data taken as a whole. While agency scientists conceded that there were weaknesses in individual studies, they saw these defects as less significant than the consistency across studies. The agency's position was concisely summed up by Theodore Farber of the Hazard Evaluation Division of OTS: "one cannot ignore the fact that there were about seven studies here on the parent and the metabolite, and that there appears to be a common thread that runs through most of these studies, in terms of a certain tumor being seen, or tumors being seen."[102] This "common thread," EPA concluded, indicated a real oncogenic (tumor-producing) effect associated with both daminozide and the metabolite UDMH.

Uniroyal's strategy, by contrast, was to discredit each of the studies individually, so as to undercut any possible support for the contention that daminozide and UDMH were carcinogens. To shake EPA's case, Uniroyal advanced familiar arguments relating to the interpretation of rodent bioassays and short-term tests. The company's experts claimed, for instance, that daminozide itself was nonmutagenic and UDMH only weakly so, if at all—facts that tended, if true, to weaken the hypothesis of carcinogenicity.[103] With respect to the Air Force Study, Uniroyal's position was that the route of exposure (inhalation) rendered the results irrelevant for use in estimating risks from ingestion.[104] The NCI study was likewise discounted for purposes of human risk assessment because of irregularities in the dose-response curve.[105] The most aggressive scientific criticism, however, was reserved for the Toth studies. Uniroyal experts claimed that the design, conduct and interpretation of these studies were flawed in so many respects that they could not reliably be used as a basis for drawing any conclusions at all about the carcinogenicity of daminozide or UDMH.

Among the failings of the Toth studies, the most important in Uniroyal's view was that the maximum tolerated dose (MTD) had been exceeded in several instances. This gave rise to the suspicion that the observed tumors were caused by secondary toxicity rather than the oncogenicity of daminozide and UDMH.[106] Uniroyal experts also criticized the Toth group for failure to observe good laboratory practices, for example, through inadequate monitoring of dosage and failure to account for the fact that test animals were treated with antibiotics.[107] An even more serious charge was that the Toth group had compared tumor incidences in the test animals with data from historical rather than concurrent controls.[108] In one case, contemporaneous data were unavailable, apparently because the researchers lacked funds for preparing tissue slides from the control animals.[109] Whatever the reason, however, Uniroyal representatives claimed that the absence of genuinely comparable data from contemporaneous controls rendered Toth's observations useless, particularly as a basis for human risk assessment.

To establish these points more firmly, Uniroyal resorted to boundary work, dismissing the Toth studies as material that did not deserve to be regarded as "science." Gerald P. Schoenig, Uniroyal's principal toxicological consultant, developed this theme in detail. Schoenig told the SAP that he had reviewed the Toth studies not for technical violations of good laboratory practices, but from "the point of view of basic science."[110] He suggested that there were certain "essential elements" that a laboratory experiment involving research animals should contain in order to be usable for human risk assessment,[111] and that the Toth studies fell short of meeting these standards. Indeed, the only respect in which the studies conformed to Schoenig's scientific paradigm was in their use of an adequate number of animals per test group.[112] Other Uniroyal spokesmen echoed Schoenig's conclusions about the status of Toth's work. Citing the use of excessive doses and irregular controls, Vesselinovitch observed that "The studies per se fall short in their design, execution, and justified interpretation of data relevant to [a] scientific conducted screen for potential carcinogenicity."[113] Dr. Albert Kolbye, a former FDA toxicologist, agreed: "Again, the Toth studies, to me, are just—they are not even within consideration. I don't even really want to call it science."[114]

These efforts to place the Toth studies outside the boundaries of science proved extraordinarily persuasive to the SAP. Predisposed by disciplinary training to impose strict standards of acceptability on toxicological data, the panelists easily fell in with the rhetoric of "good science" and dismissed EPA's efforts to assess the risks of daminozide. Some of their comments recalled the frustrations expressed earlier about EDB:

> I would like to ask that, in view of this desire to get better science into risk assessment, whatever has induced the Agency to suddenly dash into this whole issue with data that frankly just don't seem to exist. This is not science. What's the big rush on this at the present time? Why do we have to look at audits and audits and audits on data that doesn't come close to GLPs?[115]

Other panelists, too, were disturbed at what they saw as a rush to judgment on EPA's part. One member asked why, if the agency was prepared to wait one or two years to complete the cancellation process, it could not afford to wait three or four years to obtain more complete toxicological data.[116]

EPA representatives responded defensively to these concerns. One official noted that waiting for new studies would actually draw out the period of inaction to five or six years, since additional time would be needed to evaluate the results of any new studies commissioned from industry.[117] Another emphasized that the judgment whether or not to risk continued public exposure should properly be made by EPA rather than the advisory committee. Paul Lapsley, chief of the Special Review Branch, tried to redraw the boundaries so as to limit SAP's scope of review:

> I think it is important to point out the issue before the Panel is not whether risk from an additional three years of use is a significant concern. The issue that we have asked . . . the Panel's recommendation on, is whether the currently available data support the weight of the evidence conclusion that daminozide and UDMH are carcinogenic in animals, and has the potential to be carcinogenic in humans.[118]

SAP, however, was intent on imposing its own view of rationality on EPA, and this singlemindedness initially led the panel to overlook one of EPA's most important queries: were daminozide and UDMH *qualitatively* "carcinogenic in animals"?[119] This provoked an unusual altercation between SAP and OPP director Steven Schatzow,[120] who felt that the scientists were not addressing the questions put to them by the agency. Following a heated interview, the panel eventually added a sentence to the effect that the data were inadequate to support even a qualitative finding of carcinogenicity. But the panel also went on to criticize the agency's overall conduct of the daminozide review, implicitly finding fault with the sense of emergency conveyed by PD 2/3/4.[121]

This assessment effectively ruled out the possibility of regulating Alar on the basis of the existing information, for OPP concluded that it would be impractical to go against SAP's judgment.[122] Over the next three months, the agency met with manufacturers and users of Alar to map out an alternative strategy. In January 1986 EPA announced its intention to permit continued use of daminozide pending Uniroyal's submission of new toxicity and residue data,[123] and by late February the company agreed to conduct most of the requested studies.

Delayed Vindication

Although scientific advice put a temporary halt to federal regulatory activity, it paved the way to unexpected political ferment over daminozide. In May 1986 Massachusetts and Maine took steps to reduce and eventually eliminate daminozide residues in food sold in those states. At the same time, to Uniroyal's dismay, the Natural Resources Defense Council and Public Citizen, the Nader health research group, persuaded several supermarket chains (Safeway, Kroger, Giant, and Grand Union) to discontinue the sale of apples treated with Alar. A company spokesman deplored these actions, saying, "We feel that Mr. Nader and the Natural Resources Defense Council are scaring consumers without any scientific basis."[124] Joined by Public Citizen and the states of New York and Maine, NRDC filed a petition with EPA to establish zero tolerances for daminozide. Later, with Ralph Nader as primary plaintiff, the petitioners sued the agency in federal court for denying their request.[125]

Both in their petition to EPA and in their brief to Court of Appeals for the Ninth Circuit, the public interest groups reemphasized OPP's original reasons for accepting the existing cancer studies. Daminozide and UDMH, they argued, had been shown to cause tumors in test animals in multiple tests, at multiple sites, and through multiple routes of exposure. Even if EPA's quantitative risk estimate was off by two orders of magnitude, they noted, the risk would still be high enough to preclude leaving the product on the market for four more years. Under the circumstances, the petitioners argued, it was "arbitrary and capricious" for EPA to change its position simply on the basis of a meeting with the SAP, a meeting which not only failed to raise any new scientific issues but unlawfully shifted the burden of proving that Alar was safe from the manufacturer to the agency. In October 1988 the Ninth Circuit ruled against the plaintiffs, but without reaching the merits of their arguments.[126] The court held that the plaintiffs had sought judicial review on improper grounds: they should either have filed objections with the agency for denying their petition or have appealed the daminozide regulation issued by EPA only a few days later.

In a curious reversal of earlier patterns, science rather than law came to the petititoners' rescue. Early in 1989 Uniroyal announced that the now completed daminozide bioassays showed negative results in rats, but a nonstatistically significant increase in malignant and benign blood vessel tumors in mice. Moreover, preliminary results from the still incomplete UDMH studies showed a confirmatory positive carcinogenic response in mice at high doses.[127] Based on these interim results, EPA calculated an excess lifetime cancer risk of 5 in a million to adults and of 9 in a million to children exposed for 18 months. Acting Administrator John A. Moore considered these findings sufficiently disquieting to justify accelerating the daminozide cancellation proceedings. In a letter to the International Apple Institute, Moore noted that the process could still cover a period of several years and encouraged industry to take voluntary action to reduce public exposure.[128] In the next few weeks, however, the power to control policy on Alar slipped dramatically out of EPA's hands.

While Uniroyal was generating new data on daminozide and UDMH, NRDC had launched a two-year study, peer-reviewed by an "independent panel," of the risks to children from a variety of widely used chemicals, including Alar.[129] The report assessed the cancer risk from Alar as several hundred times higher than estimated by EPA. In a remarkably effective display of scientific self-help, NRDC persuaded the CBS Television news program "60 Minutes" to cover the release of its report. CBS reporter Ed Bradley described Alar as "the most potent cancer-causing agent in our food supply" and represented this as "the conclusion of a number of scientific experts."[130] The ensuing public furor overwhelmed attempts by both industry and EPA to impugn the credibility of NRDC's report. Charges that NRDC's risk estimates were based on studies discredited by the SAP had little discernible impact on public opinion. Consumers, it seemed, saw no reason why NRDC's expertise was less to be trusted than that of EPA or the SAP, let alone the agriculture lobby. Uniroyal bowed to public and governmental pressure by deciding to discontinue production of Alar. OPP's initial policy thus prevailed, but through a process that transformed science into a servant of politics.

For the SAP, in particular, a moment of deeper humiliation came in May 1989, when two senators, Henry Reid (D-Nevada) and Joseph Lieberman (D-Connecticut), charged that EPA's pesticide program and its advisory panel were "riddled with pro-industry bias."[131] Senatorial attention was drawn especially to the actions of Christopher Wilkinson and Wendell Kilgore, two of SAP's most respected toxicologists. At the root of Wilkinson's difficulties was his decision, following his departure from the SAP, to represent Uniroyal on a technical issue related to Alar. Wilkinson helped the company advance the argument that the new cancer bioassays required by EPA were being conducted at doses exceeding the MTD, a practice that was likely to cause premature death and other problems in the test animals. This behavior, in the view of some observers, violated provisions of the Ethics in Government Act, which permanently bans former federal employees from representing clients on any "particular matter" that they handled while in government service. Though Wilkinson denied any wrongdoing, the incident showed that the mantle of expert adviser would give scientists small protection against charges of bias and capture.

A Fragmentation of Authority

When Congress established the SAP, it hoped to endow EPA's pesticide program with an advisory body of such integrity and expertise that its word would count as final on all issues of regulatory science. The SAP clearly did not live up to these expectations in the Alar controversy. The question of risk was ultimately decided in a nonscientific forum and in a fashion that ignored most of the subtleties and complexities of using toxicological evidence to predict risk. By the time NRDC's report was aired on "60 Minutes," the SAP had become largely irrelevant. But the panel's loss of control over the scientific debate was not entirely accidental; the Alar case simply brought to light a pathology that was latent in the relationships among the panel, the agency, and concerned interest groups.

Proceedings before the SAP were so structured as to highlight rather than blur the socially constructed character of the risk estimates produced by EPA and the agrochemical industry. The conflicts over EDB, dicofol, and Alar at SAP meetings can be seen at one level as classic studies in the social construction of science.[132] In each case, OPP and industry used different conventions to determine whether the existing data were scientifically acceptable and a suitable basis for risk assessment. Their scientific choices, moreover, were consistent with their overarching institutional imperatives. OPP was under pressure from Congress to establish reasonable priorities and to speed up the reregistration of substances approved prior to 1972. Accordingly, OPP officials were favorably disposed to accepting existing studies, especially when these appeared to satisy criteria that EPA had considered sufficient in the past for qualitative and quantitative assessments of risk.

By contrast, industry's interests obviously favored greater caution in interpreting risk data. An insistence that OPP should wait for better evidence worked to the manufacturer's commercial advantage, since challenged products would remain on the market pending further study. Uniroyal's "divide and conquer" approach to reading the daminozide studies appears quite reasonable in the light of this analysis. By focusing on the weaknesses of each study, Uniroyal experts undermined the basis for a ban on daminozide, established the need for better data, and cemented the case for delaying regulation.

The SAP members approached their review of the scientific data from still another professional perspective. Institutionally, their role was to ensure that "good science" was upheld in the regulatory process. Thus the panelists—in particular the toxicologists—felt no qualms about insisting that each animal study should either meet state-of-the-art standards for bioassays or be rejected as inadequate. By holding the studies to the norms of current research in the field, the SAP members also satisfied their professional need to draw bright lines between "good" and "bad" science. But in both the dicofol and daminozide cases, the definition of "state-of-the-art" that SAP accepted was put forth by industry experts. This coincidence of position did not cast doubt on SAP's impartiality as long as the scientific debate remained internalized in panel meetings. In the Alar controversy, however, the debate went public, and SAP's agreement with Uniroyal's reading of the Toth studies proved sufficiently questionable to provoke a backlash in favor of the alternative reading provided by NRDC.

The cases discussed in this chapter underscore a serious deficiency in the SAP advisory process: the absence of mechanisms for reconciling alternative constructions of science. Industry's practice of presenting a battery of carefully primed scientific advocates realized earlier fears about the SAP by converting panel meetings into adversarial exercises where EPA was invariably cast as the weaker party. Contests between registrants and the agency were generically unequal because industry's experts had more impressive professional credentials than OPP's. SAP meetings, moreover, offered little opportunity for OPP to rebut the arguments raised by industry, and public interest groups did not make noticeable efforts to redress the balance. Yet, because the proceedings lacked the safeguards of formal administrative hearings, even a well-founded endorsement of industry's position by the SAP was apt to be viewed by public interest groups as politics capturing science.

Together, the three cases illustrate the political malleability of peer review. Although the SAP was not appointed by the White House as some critics charged, its credibility as a reviewing body was weakened when it was reconvened in the early 1980s by an administration that was apparently seeking to control EPA's scientific advisory network (see Chapter 5). Political interest groups, too, learned to use peer review in strategic ways. In the dicofol proceedings, Rohm and Haas enhanced the qualifications of its experts through carefully orchestrated peer review. Similarly, in the daminozide case, NRDC strengthened its claim to scientific authority (at least in the public's eye) by submitting its report to a panel of "independent" experts for peer review.[133]

How, finally, should one assess the impact of SAP's advice on EPA's decisionmaking? In the aftermath of the panel's initial evaluation of Alar, Steven Schatzow, former pesticide program head, suggested that scientific review has effectively established a double standard for pesticides.[134] Because of SAP's demand for "good science," it is harder to ban pesticides already on the market than to deny registrations to new products. In particular, SAP has grown increasingly skeptical about the appropriateness of regulating a commercially significant product solely on the basis of animal studies. At least in EPA's pesticide program, then, regulatory peer review seems to have exacerbated what some critics regard as an unfortunate tendency on the part of U.S. risk managers to favor relatively dangerous "old" risks over possibly safer "new" ones.[135] It has also created incentives for interested parties to remove the risk debate to the media and the corridors of politics, where it can be conducted under different and more flexible ground rules from those acceptable to science.

8

FDA's Advisory Network

The Food and Drug Administration is in some respects even more politically exposed than EPA, and its need for scientific legitimacy is no less urgent. FDA's mandate covers issues of the liveliest concern to the public: the safety and effectiveness of pharmaceuticals, the purity and wholesomeness of food, and the safety of cosmetics and ingredients added to food. Regulatory missteps by the agency can generate tremendous political shockwaves, frequently precipitating a legislative response. Thus, the discovery that an unreviewed and unapproved drug, sulfanilamide, had caused more than a hundred deaths led in 1938 to the adoption of a premarket approval system for new drugs, and the thalidomide tragedy provided impetus for the requirement that drugs should be proved effective as well as safe before being marketed.[1] More recently, FDA has been accused of dragging its heels on new drugs, so that beneficial drugs are introduced much more slowly in the United States than in other countries, creating the so-called drug lag. Claims of scientific expertise have not proved particularly effective against charges of timidity and bureaucratic inertia, especially when FDA restricts access to products that the public urgently wants, such as azidothymidine (AZT) for AIDS or tissue plasminogen activator (TPA) for heart disease.

Starting in the 1970s, FDA's regulation of food additives also became politically sensitive, though controversies in this area were more sporadic and unpredictable than in the drug program. FDA-watchers were appalled at the agency's attempt in 1976 to ban saccharin, then the only artificial sweetener on the market, on the basis of what appeared to be a very slight risk to human health. FDA's subsequent abortive effort to eliminate nitrites reinforced concerns about overregulation and increased the demand for a more balanced approach to risk assessment, accompanied by regulatory peer review.

During the Reagan administration, by contrast, the agency drew criticism from Congress and the press because of its glacial pace of decisionmaking on a group of potentially carcinogenic color additives. Though these disputes carried confused messages for FDA, they reflected, at bottom, a pervasive public dissatisfaction with the implementation of the Delaney clause, the controversial anticancer amendment to the Federal Food, Drug, and Cosmetic Act (FDCA).[2] As we shall see, the agency's bold attempt to redefine the provision in keeping with current knowledge about cancer risks also failed, illustrating the pitfalls of using scientific advice to address a fundamentally political and legal problem.

Over almost a century of involvement in food and drug regulation, FDA has built up an enviable record of deferential treatment from the courts, which gives it a decided advantage over newer science policy agencies such as CPSC, OSHA, and EPA. Yet the agency has experienced chronic trouble attracting and holding highly qualified physicians and research scientists, and it is perpetually understaffed.[3] The formidable array of expert advisory committees associated with the agency since the late 1960s testifies to its recognition that legitimation from the independent scientific community is indispensable to the success of its regulatory programs. When these external advisers work in harmony with the agency's in-house experts, the resulting policy decisions seldom need any further defense on scientific, legal, or political grounds.

The role of advisory committees in FDA's decisionmaking, however, is not invariably so productive. In turning to the scientific community for advice, the agency runs the risk of having its own construction of regulatory science challenged (or deconstructed) rather than validated. When this happens, FDA's expert authority unravels, and the agency comes under fire for failing to abide by the judgments of its external referees. Even when the agency concurs with its advisers, it is frequently accused of needlessly relying on advice as an excuse to postpone decisions. The factors that lead to such negative appraisals can best be understood through an analysis of specific cases. The culture of advice-giving at FDA is first explored through the agency's historical experiences with advisory committees in its drug-approval program. Three more recent proceedings are then examined: the development of guidelines for the clinical evaluation of antiarrhythmic drugs, the regulation of carcinogenic color additives, and the partial ban on sulfites in food.

The Scientific Evaluation of Drugs

As FDA's mission and powers have grown, so have the number and types of advisory committees the agency employs. As of 1986, there were thirty-nine standing committees and panels associated with the Commissioner's Office, the National Center for Toxicological Research, and the Centers for Drugs and Biologics, Devices and Radiological Health, and Veterinary Medicine.[4] Some of FDA's advisory committees are mandatory, such as the Device Review Panels required under the medical-device provisions of the FDCA. For the most part, however, the agency relies on voluntarily established standing committees or solicits advice through ad hoc mechanisms consistent with its overall mission.

Beginning in the 1960s, the Bureau of Drugs has routinely sought outside advice on a variety of scientific matters related to the screening and marketing of pharmaceutical products. Consultation with advisory committees has become the rule in connection with Investigational New Drug Applications (INDs) and New Drug Applications (NDAs). Committee approval is also sought for new uses of marketed drugs and in reviewing classes of products when the agency believes there is a scientific basis for reevaluation. Occasionally, FDA also refers emergency decisions concerning already marketed drugs to advisory committees, as for example in the controversy discussed below over remarketing the painkiller Zomax.

The Department of Health and Human Services at one time regularly delegated the responsibility for selecting these advisory committees to the FDA commissioner, who further subdelegated this authority to the appropriate division level. Beginning with Joseph Califano's tenure as Secretary of Health, Education and Welfare, however, the Department began exercising more control over committee appointments, allegedly in the interests of securing greater uniformity and quality.[5] The pattern intensified during the Reagan administration, with the result that the influential Cardiovascular and Renal Drugs Advisory Committee (hereafter referred to as the Cardio-Renal Committee) was left as the only one still appointed by an FDA division. Few within the agency regard this change as salutary. The Department's involvement added another layer of bureaucracy to an already burdensome process and exposed FDA division staffs to increased risk of political pressure. Advisory committee appointments are regarded as professional plums in the medical community, opening the door to lucrative private consulting opportunities. Accordingly, physicians are not averse to political maneuvering in order to secure a nomination to an FDA committee. Agency personnel are not convinced that the benefits of removing appointments to a higher administrative echelon offset these considerable disadvantages.[6]

Normally little noticed by the public, FDA's interactions with its drug advisory panels received extraordinary attention in 1974, at a series of hearings conducted by Representative Fountain of North Carolina, one of the agency's most formidable critics in Congress.[7] The Fountain subcommittee accused FDA of numerous misuses of the advisory process, leading to alleged laxness in the evaluation of some new drugs. By the decade's end, however, overzealous regulation became the primary charge leveled against the agency. Based largely on comparisons of approvals of new drugs in the United Kingdom and the United States, drug industry advocates and economic analysts argued that FDA's cumbersome procedures were stifling innovation and depriving the public of beneficial therapies already available in other leading pharmaceutical-producing nations.[8] In contrast to Congress, these critics proved to be enthusiastic supporters of FDA's advisory system. Their views coincided with those of former FDA Commissioner Alexander Schmidt, who testified to the Fountain subcommittee that outside experts shore up the agency in areas where it is technically deficient or unduly cautious, thereby permitting prompter resolution of matters that FDA could not hope to deal with decisively on its own.[9]

Safety and Efficacy Reviews

In order to put a new drug on the U.S. market, the producer must establish to FDA's satisfaction that it is both safe and effective. To satisfy the efficacy requirement, which was added to the FDCA by the 1962 Kefauver-Harris Amendments, the manufacturer must provide "substantial evidence" that the drug is effective. As defined by the Act, this means "evidence consisting of adequate and well-controlled investigations, including clinical investigations, . . . on the basis of which it could fairly and responsibly be calculated . . . that the drug will have the effect it purports or is represented to have."[10] The primary duty of advisory committees attached to the drug evaluation program has been to ensure that the evidence on safety and efficacy complies with statutory requirements. FDA and its advisers have not always agreed, however, on the correct way to interpret the substantial evidence test.

Evidence and Interpretation

An example of FDA's decisionmaking that particularly incensed the Fountain subcommittee was the agency's apparent disregard of scientific advice in approving the antiarrhythmic drug propranolol (trade named Inderal) for use in treating angina pectoris. FDA staff had concluded as early as 1968 that propranolol might be effective against angina. Until early 1973, however, the agency did not deem the evidence sufficient to satisfy its scientific advisers or to comply with applicable legal and regulatory requirements. In April of that year, FDA called upon its Cardio-Vascular and Renal Drugs Advisory Committee to review thirty-three studies relating to the proposed new use of propranolol. Members were asked to consider three options: to approve the drug outright for angina; to recommend no changes in the existing package insert; to change the package insert so as to provide guidance for physicians who were already prescribing propranolol for angina patients.[11]

The medical and scientific picture related to propranolol had by this time changed very considerably from when it was originally approved for arrhythmia. By 1973 the drug's major actual use was for angina, a use not officially sanctioned by FDA.[12] The manufacturer, Ayerst Laboratories, was conducting clinical trials designed to establish the efficacy of propranolol for long-term treatment of angina, but these studies were not expected to be completed for another two years. Enough additional studies had accumulated in the literature for FDA's own medical officials to believe that there was an adequate scientific basis for immediate approval of the new use. In the meantime, however, the agency faced a widening credibility gap as clinical practice outstripped its approval process. Consultation with the advisory committee, the agency hoped, would confirm the agency's internal scientific judgment and would permit a quicker resolution than if the agency waited for Ayerst to submit an NDA according to more usual practices.

The risks in this strategy became apparent only after the advisory committee completed its deliberations on the drug using standards that seemed too rigid to FDA. The major difficulty from FDA's point of view was the committee's refusal to review the available studies en masse, as the agency's own experts were prepared to do. Instead, the physicians on the committee adopted a case-by-case approach to reviewing the methods and conclusions of each study. Predictably, this style of scrutiny persuaded the panelists that none of the studies was methodologically sound and that none met the standards of adequacy imposed by the FDCA and the implementing regulations.[13] The committee then divided on the question of what should be done with propranolol. Given the inadequacies of the scientific data base, no member was prepared to push for outright approval of the new use. One panelist argued that there should be no change at all in the official status quo, while the remaining four members opted for a less extreme position, namely, that the package insert should be rewritten without giving formal approval for the new use. This action reflected the majority's professional judgment that the drug was "probably effective," although no individual study could be certified as scientifically adequate and well controlled.[14]

The advisory committee's decision left the agency with two choices, neither of which was ideal: either to go along with the limited proposal actually approved by the committee or to proceed with the full-approval plan despite the committee's hesitations about the adequacy of the data. FDA chose the latter option. The director of the Division of Cardio-Renal Drug Products tried to characterize the evidence in the strongest possible light in the "summary basis for approval": "Although every study submitted did not in each instance meet all criteria, as a group they do provide evidence of a well-controlled quality to support safety and efficacy."[15] Of course, this conclusion did not jibe with the actual findings and recommendations of FDA's scientific advisers. A member of the advisory committee[16] and a consultant employed by the Fountain subcommittee charged the agency not only with flouting its advisers' judgment but with violating the legal requirement that approval for new uses be conditioned on the availability of at least two adequate and well-controlled clinical trials.[17]

FDA's self-defense, vigorously conducted by Bureau of Drugs director Dr. Richard Crout and general counsel Peter Hutt, was remarkable for its adroit use of boundary work. Both Crout and Hutt maintained, in the first place, that the evaluation of scientific evidence by the advisory committee should be carried out according to standards of scientific rather than legal judgment. Hutt, indeed, insisted forcefully, though rather misleadingly, that the determination of whether the propranolol studies added up to "substantial evidence" was not a legal but a scientific or medical exercise:

> I would say the last person in the world who I would want to decide whether propranolol or any other drug is supported by substantial evidence would be a lawyer. I would want that to be determined by a physician qualified by training and experience to make those types of determinations.[18]

Seconding Hutt, Crout also emphasized that the committee had the power to review the thirty-three studies as a whole and to conclude, if necessary, that together they constituted substantial evidence.[19] It was within the scope of the committee's medical judgment, Crout implied, to decide that the existing data established propranolol's efficacy, even if the regulatory requirement of two specific well-controlled trials could not be literally satisfied.

At the same time, taking virtually the opposite tack, Crout reminded the legislators that the authority to approve a drug was lawfully delegated to the agency, so that the ultimate decision on how to act rested with him alone: "And I am the responsible individual for those decisions and a responsible person takes advice. I take the advice of our own staff and we take advice of our advisory committees, and you use what knowledge you have yourself in such a decision."[20] This contorted logic permitted Crout on the one hand to absolve the committee of the need to follow the strict letter of the law in evaluating the evidence (the question was scientific, not legal); on the other hand, he absolved himself, and by extension the agency, of the need to abide by the committee's assessment if it failed to meet FDA's objectives (the decision was policy, not science).

FDA's reading of its responsibilities in this case won influential support outside Congress. William Wardell, a leading commentator on federal drug regulatory policy and a vocal critic of the drug lag, applauded FDA's decision on propranolol as a manifestation of reason and sound administrative judgment. Not surprisingly, he was highly critical of the Fountain subcommittee:

> A group of 13 studies—which obviously showed the efficacy of the drug—was deemed by FDA to satisfy the law's substantial evidence requirement. FDA was then unfairly harassed for several days at hearings of a Congressional advisory committee which, using biased and erroneous interpretations from an inappropriately chosen consultant, alleged that every one of the 13 studies had fatal defects (an

> inaccurate assertion) and that FDA had therefore broken the law in approving propranolol for angina since the requisite two well-controlled studies did not exist! This is a bizarre case of political pharmacology, as well as an unwarranted criticism of the Agency . . . (citations omitted.)[21]

But the label "political pharmacology" could as easily have been affixed to FDA's decision to override a cautious advisory-committee recommendation. This at any rate seemed to be the perception of members of the Fountain subcommittee.

Using advisory committees to compensate for gaps in the scientific dossier does not always shield FDA decisionmakers against political embarrassment, even when the committee agrees with the agency. FDA discovered this in the case of Zomax, the first nonsteroidal anti-inflammatory drug approved for treating mild to moderately severe pain.[22] Within months of its approval, reports of allergic reactions began appearing in the literature, and in March 1983 the manufacturer voluntarily withdrew the drug following five reported fatalities resulting from its use.[23] In August of that year, FDA approached its Arthritis Advisory Committee with a proposal to "reposition" Zomax for the relief of chronic pain, such as cancer pain, rather than for the intermittent use that appeared responsible for most of the allergic reactions. After hearing presentations by the agency and the manufacturers, the committee recommended remarketing, on condition that the manufacturer would carry out studies to prove that there was a population of patients for whom the benefits of the drug outweighed its risks.[24]

An investigative House subcommittee, headed by Congressman Ted Weiss, again chastised FDA for violating its own rules and practices for consulting outside advisers. It alleged, in particular, that FDA should have identified the class of patients for whom Zomax would be more beneficial than other drugs in its class *before* submitting the issue to the committee.[25] The report also suggested that the agency should not have gone to the committee without more scientific evidence to back up the repositioning proposal. As in the case of propranolol, the congressional inquiry focused on apparent legal and procedural irregularities in the relations between the agency and its advisory committee. At a deeper level, however, the issue was clearly political. The Weiss committee disagreed with FDA over the level of risk the public should tolerate from Zomax. Criticism of the use of scientific advice was merely a surrogate for the legislature's real complaint that FDA should have demanded more stringent proof of safety and efficacy before remarketing a possibly lethal drug.

Although consultation with advisory committees backfired in the propranolol and Zomax cases, such interactions can in principle be extremely helpful to FDA when it is confronted with close risk-benefit decisions on new drugs. Turning to a committee for advice subtly recasts an inherently political judgment about the acceptability of risk as a scientific judgment about the acceptability of evidence. FDA successfully exploited this kind of frame-shifting in the case of Triazure, a drug of proven effectiveness for treating severe psoriasis, but posing a serious risk of thromboembolism to some patients.[26] The agency's initial decision to permit the marketing of Triazure was supported by four out of five members of the Dermatology Advisory Committee, but another committee composed primarily of epidemiologists recommended that FDA await additional premarket studies. Director Crout chose to rely on the Dermatology Committee, emphasizing its technical credentials; it was composed, he said, of "medical experts experienced in the use of drugs and in the evaluation of the range of evidence available on a particular drug."[27] This bow to the committee's expertise camouflaged FDA's fundamentally political decision to side with the scientists who favored marketing. A supportive observer characterized the events as a "judicious use of advisory committees and a courageous and wise decision on the part of the Bureau of Drugs."[28] Ironically, the same individual interpreted FDA's subsequent withdrawal of Triazure as an indefensible bow to political pressure rather than a justified shift to a more precautionary reading of ambiguous evidence.

Antiarrhythmic Drugs: Evolving a New Philosophy

Many at FDA believe that outside scientists can play an even more constructive part in developing policies for classes of therapeutically important compounds than for single products. The agency's reconsideration of guidelines for conducting clinical trials for antiarrhythmic drugs provides strong support for this view. Input from the Cardio-Renal Committee proved to be a crucial factor both in initiating these proceedings and in helping the agency respond to unexpected medical findings concerning these products.

FDA's policy on antiarrhythmic drugs followed an erratic course after the approval of propranolol for this indication in 1968. From

1968 to 1976 no new drugs were cleared for treatment of arrhythmia, except for lidocaine, which was "belatedly approved" as an antiarrhythmic in 1970 following a period of resistance by the Cardio-Renal Division.[29] This prolonged inaction angered critics of FDA, who pointed to the ever-increasing lag between Britain and the United States with respect to antiarrhythmics. In a 1978 article, William Wardell criticized FDA for its retrograde regulatory philosophy, citing with particular disdain a medical official who favored limiting the number of new products in order to avoid creating a "therapeutic Tower of Babel."[30] Restricting the U.S. market in antiarrhythmics was felt to be particularly unfortunate because these drugs frequently induce serious side-effects and it is thus important for physicians to have a reasonable number of alternatives available. Spurred by such criticism, the introduction of new antiarrhythmic drugs speeded up markedly in the Reagan administration. According to one agency official, six such products were approved in 1984 and 1985, compared with only four in the previous decade.[31] After considering each of these new drugs case by case, the Cardio-Renal Committee was convinced that the time had come for a total reevaluation of the field and urged the agency to develop new guidelines for clinical trials of antiarrhythmic drugs awaiting approval by FDA.

Agency officials describe the review conducted by the advisory committee in 1985 and 1986 as involving issues that were part regulatory and part philosophical.[32] All antiarrhythmic drugs are relatively toxic; moreover, in the view of many experts, their efficacy had not been convincingly demonstrated. While these drugs are effective in reducing the frequency of abnormal heartbeats, evidence that this corresponds to a reduction in the risk of life-threatening arrhythmias appeared less compelling. Yet with as many as 25–30 new drugs under development, the need to ascertain their effectiveness as completely as possible was clearly pressing. Issues that the Cardio-Renal Committee and FDA staff felt were ripe for reconsideration included the following: who should use the drug, how should the hazards of use be communicated through labeling, and how should clinical trials be conducted in order to establish the case for efficacy more definitively?

Participants in the July 1985 Cardio-Renal Committee meeting began with certain assumptions in common about the nature of the task ahead. There was general agreement that the testing done on recently approved antiarrhythmic drugs had been done mainly on populations for which there was little or no clinical indication that particular drugs would be effective.[33] Patient populations had been selected primarily on the basis of the nature of the arrhythmia rather than the characteristics of the patient in whom the arrhythmia occurred. Because of their toxic side-effects, however, most drugs of this type were believed likely to work only for particular subpopulations of patients. The committee agreed that it would be useful for physicians to have better guidance on the nature of these subpopulations for each drug, so that prescriptions could be tailored to the groups that would benefit most from each therapy. Yet the testing currently undertaken by manufacturers was obviously not sensitive enough to produce the data for this type of improved guidance.

While committee members concurred that patient populations for clinical trials should be better characterized, they disagreed about the way these subgroups should be defined. In particular, it was clear to the whole committee that refining the test guidelines to address narrower patient groupings would impose additional costs on manufacturers during the investigative phase and would potentially constrict postapproval sales.[34] The revised guidelines would require sponsors of each drug to collect information about patient characteristics prior to approval, rather than through postmarket reporting by physicians. Differences emerged on the committee as to how much of the information burden it was appropriate to shift from the postapproval to the preapproval stage. Concerns about the costs of running more differentiated trials, as well as the statistical feasibility of working with smaller (because more specifically characterized) patient groups, led committee members to adopt rather different positions.[35]

Typologically, the issues confronted by the committee were similar to issues that often prove divisive in regulatory proceedings. How much information should manufacturers produce? Is testing technically feasible? What costs can appropriately be imposed on producers in order to ensure the development of safer products? But because these questions were raised in the context of informal expert deliberations they created surprisingly little conflict.

The discussions carried on by the Cardio-Renal Committee were significantly different in character from SAP meetings in EPA's pesticide program. Neither the agency nor other interested parties made formal presentations, so that committee members were not cast in the role of judges or interrogators. At the start of the first public meeting, each committee member received a set of draft guidelines prepared by a three-member subcommittee and internally reviewed by the committee chairman and an FDA expert to identify the issues on which consensus should be developed.[36] The immediate aim was to discuss these issues in a way that would permit the subcommittee to prepare a revised draft for consideration by the full committee at its next meeting. This meant, in effect, that the committee was acting as a peer-review body for a subset of its own members. Hence, the potential for adversarial relations with agency staff was kept to a minimum.

A noteworthy feature of the Cardio-Renal Committee's meetings was the role of those representing industry and FDA. As the head of the Division of Cardio-Renal Drug Products, Raymond Lipicky played an active part both in the formulation of the new guidelines and in the committee's subsequent deliberations. His position, however, was not that of an advocate for a developed agency policy. Rather, Lipicky's main contribution was to steer the discussion on a still-fluid issue of regulatory science in directions that were most useful for FDA. Industry's participation, as well, was more as consultant than as interested party. For example, Dr. Gary Jensko, medical director at Riker, intervened in the discussion a number of times, but mostly in response to specific queries.[37] The committee treated drug company representatives like Jensko mainly as specialized informants on the practical impact of the proposed guidelines. In sum, the absence of a hard-and-fast agency position, the consultative status of both agency and industry representatives, the focus on a general product class, and the nonadversarial format of the proceedings gave the Cardio-Renal Committee's meeting on antiarrhythmics more the character of a European than an American exercise in using scientific advice.[38]

One effect of the 1985–86 proceedings before the Cardio-Renal Committee was to validate FDA's earlier concern about the random "noise" surrounding these products. In 1985 as in 1978, it seemed fair to say that antiarrhythmics as a class

> were complex, unpredictable and paradoxical (i.e., in some circumstances will produce the condition sought to be treated); that there were a variety of agents . . . ; that "few, if any, physicians are able to be fully knowledgeable concerning the complex pharmacologic variations and characteristics of all these agents," [and] that unassessable drug effects result if these agents are used concurrently or with other drugs.[39]

FDA's response to this complexity before the committee intervened left much to be desired: first, regulatory paralysis, and then, under rising criticism, a perhaps ill-considered zeal in approving a vastly expanded, but inadequately tested, array of new products. With the Cardio-Renal Committee's backing, FDA was empowered to move toward a more balanced position, one that might otherwise have proved too politically costly.

The committee's support also proved invaluable to FDA as new clinical findings on two widely used antiarrhythmics, encainide and flecainide, unexpectedly raised the level of controversy surrounding the entire class of products. The Cardiac Arrhythmia Suppression Trial (CAST), a study designed to test the efficacy of antiarrhythmic therapy in a subclass of patients with mild ventricular arrhythmia, had selected both drugs for evaluation after it was shown that they were effective in suppressing the irregular heartbeats. But in August 1989 a special report in the *New England Journal of Medicine* announced that the CAST patients treated with these drugs had a higher rate of death than patients assigned to placebo.[40] The investigators accordingly discontinued the part of the trial involving encainide and flecainide and recommended that neither drug be used to treat the subset of patients studied in CAST.

Although these results prompted at least two drug companies to change the labeling for encainide and flecainide, they also unleashed a flood of questions within the medical community about the proper way to interpret the CAST report.[41] Following the usual pattern of experimenters' regress, one set of queries focused on the results observed in the control group. If the death rates in the placebo group were too low, as some suggested, then the CAST population might have represented an artificially healthy subset of arrhythmia patients, tending in turn to exaggerate the relative risk between patients on active drug therapy and those on placebo. A distortion of this kind would have weakened the results of the study, both scientifically and as a basis for policy. A second, and from FDA's viewpoint more significant, set of questions related to the implications of CAST for labeling encainide and flecainide in particular, and for antiarrhythmic drug therapy in general.

Both types of issues were presented to the Cardio-Renal Committee at a special meeting in October 1989. After several presentations by CAST investigators, the committee considered a series of questions ranging from the relatively specific and data-oriented (did CAST find clear differences in mortality between encainide and flecainide and their placebos) to the more clinical and judgmental (would antiarrhythmic drugs benefit patients with life-threatening arrhythmias; should labeling include mention of patients not treated in CAST). Committee members indicated at several points that they were aware of two rather different roles that these questions required them to play: "clinician" as well as "scientist." For instance, Dr. Craig M. Pratt, the committee chair, exhorted his colleagues in the following terms to extrapolate beyond the CAST data:

> You need to come down to your feeling though and your judgment based on yourself as a clinician and give us these other areas. Do you think labeling should include mention of other kinds of structural heart disease? Or do you, in fact, recommend that it only be limited to CAST patients, period? And why?[42]

Echoing Pratt's unspoken dichotomy between clinician and scientist, Dr. Douglas P. Zipes, an invited guest, remarked at another point, "I think all of us, when we wear our scientist hats, would prefer a prospective, placebo-controlled trial. Obviously that is impossible."[43]

These exchanges show an advisory committee functioning in an important bridging capacity between science and public policy. The questions that FDA needed to have answered in this case were clearly "trans-scientific" in the sense that they had to extend beyond the state of scientific knowledge represented by CAST. Labeling policies, in particular, had to reach classes of patients not included in CAST and for whom prospective, controlled clinical trials were not likely to be forthcoming. By agreeing to act in two different expert capacities, the committee members continued to lend scientific authority to FDA; at the same time, they permitted the policy debate to stray usefully beyond the strict limits of CAST.

Expertise and Food Safety

FDA's relations with scientific experts on matters of food safety are less structured and more episodic than in the programs for drugs and medical devices, leading to somewhat different political conflicts and consequences. Apart from the Board of Tea Experts, an antiquated holdover from an 1897 enactment,[44] FDA has no standing committees that are exclusively concerned with the safety of foods or food additives. Instead, FDA's recent policy has been to enter into ad hoc agreements with scientific organizations when it wishes to have technical advice on such issues. Two examples of such consultation provide further insights into the interplay of science and policy in FDA's decisionmaking. In one case, FDA consulted an advisory panel following reports of extreme allergic reactions to sulfiting agents in food. In the other, scientific advice was sought to resolve a long-standing dispute about the safety of several potentially carcinogenic color additives.

The Sulfite Controversy

The term "sulfiting agent" refers generically to sulfur dioxide and five sulfite salts used in food service and processing.[45] These substances are of considerable value to the food industry because they inhibit browning of fresh fruits and vegetables; their use expanded greatly in the 1980s as restaurants began offering salad bars to a newly health-conscious public. In addition to preserving the fresh appearance of produce, sulfites are also widely used in the preparation of food intended for cooking, especially potatoes, and in beers and wines, dried fruits, and pharmaceutical products.

Sulfiting agents were identified as a possible target of regulation in the 1970s by the Select Committee on GRAS Substances (SCOGS) of the Federation of American Societies for Experimental Biology (FASEB). Additives "generally recognized as safe" (GRAS) were exempted from the 1958 Food Additives Amendment to the FDCA, which instituted a system of premarket approval for commercially added ingredients in food.[46] In 1969, however, a presidential directive urged FDA to review the status of GRAS substances to ensure that they were safe by standards of current knowledge and expert judgment.[47] FDA contracted with FASEB to carry out the necessary reviews, and in 1976 a FASEB Select Committee concluded that sulfites posed no identifiable risk to public health if used at then-current levels and in accordance with established practices.[48] It was not until 1982, however, that FDA acted on these findings through a proposal to reaffirm the GRAS status of four sulfite salts. No attempt was made at this time to consult the most up-to-date scientific literature on sulfites.

The Center for Science in the Public Interest (CSPI), a watchdog organization with a special interest in food safety, carried out its own literature survey on sulfites in response to FDA's proposed action.[49] The search uncovered several medical reports published between 1976 and 1982 documenting cases of acute sulfite sensitivity among exposed individuals. The reported allergic reactions ranged from mild skin rashes to life-threatening conditions and even occasional fatalities. Armed with these findings, CSPI petitioned the agency to ban or sharply limit the use of sulfites, but FDA opted instead for a much more limited plan to require the disclosure of sulfite use on processed and packaged foods. Simultaneously, FDA turned to FASEB for a further review of the medical literature on sulfite sensitivity.

FASEB's Role

The second FASEB review of sulfites represented in many respects a thoroughly exceptional exercise in expert consultation. To begin with, the scientific issues were highly focused, involving a well-defined risk to health, and the policy options before the agency were relatively limited in number. Second, although FDA's preliminary policy position had aroused some disagreement, the issues had not yet become publicly controversial. Finally, the review procedure, though ad hoc, was patterned on a model that FDA had used with marked success in the past, and it exploited a standing relationship between FASEB and FDA. The circumstances, in other words, favored effective use of expertise (see Marver Bernstein's analysis in Chapter 1), and review occurred in an incomparably more expeditious manner than is ordinarily achievable in the regulatory environment.

Pursuant to its contract with FDA, FASEB in early 1984 appointed an ad hoc review panel to consider how sulfiting agents should be classified in the light of newly available information relevant to their use and safety.[50] For FDA as well as FASEB, the panel's scientific credentials were of paramount importance, and the agency asked FASEB to select the panelists not only "for their qualifications, experience, and judgment," but also "for balance and breadth in the appropriate professional disciplines."[51] The project's steering committee decided that the simplest way of meeting these objectives would be to reconvene the members of the GRAS committee that had reviewed sulfites in 1976. It thus happened that 90 percent of the new ad hoc panel had prior experience with sulfites. The panel met three times over a six-month period. The first two meetings were closed to the public and were devoted to preparing an interim report, which was made available to interested parties. This document then formed the basis for a third, public meeting, at which the panel received oral comments from a dozen individuals, as well as written comments from forty-one additional groups and individuals. A final report setting forth the panel's findings and conclusions was issued in January 1985, just seven months after the panel began its work.

A Quasi-Legislative Process

As a form of scientific advice-giving, the service provided by the sulfite panel was fundamentally different from reviews conducted by EPA's scientific advisory panels. In the FASEB proceeding, it was the advisory panel, not the agency, that carried out the initial literature review and prepared the evaluation of risk. The purpose of holding an open meeting was to receive comments on the panel's own work rather than to let the panel criticize FDA's work. Unlike the SAP, for example, the sulfite panel was not required to adjudicate between the agency and the advocates or opponents of regulation, and its public meeting had much the character of a legislative hearing. Participants provided a wide range of data, interpretations, and philosophical observations that they considered relevant to the panel's final report and recommendations. Some participants, chiefly representatives of CSPI and the food industry, openly pressed for particular policy decisions on sulfites. But the proceedings also included researchers who wished to critique the tentative report or to make the panel aware of their own work on sulfite sensitivity. These intervenors appeared more interested in being professionally recognized than in influencing policy. All received equal treatment from the panel in that they spoke for a designated period of time (generally twenty minutes) and then fielded questions.

From Science to Policy

The official function of the sulfite panel was to provide FDA with a technically competent and up-to-date analysis of safety, based on the scientific literature. Ironically, however, the bulk of the discussion at the public meeting focused not on the medical data but on designing an appropriate policy response to the threat posed by sulfites. There was relatively little debate about the nature of the risk itself. No one questioned the panel's preliminary determination that sulfites do not threaten the health of the general public, nor that they do put a smaller class of sulfite-sensitive persons at potentially serious risk. There was also general agreement that more research was needed to determine what forms of sulfite in food can cause adverse reactions and how large a population might properly be regarded as sulfite-sensitive. What generated the most controversy was the panel's tentative policy conclusion that, pending the development of more information on these matters, FDA should require "clear and prominent labeling in restaurant notices where sulfited foods are being served, and labeling of foods in the marketplace."[52]

Objections to this recommendation took three forms. Speaking for consumers, CSPI argued that labeling was inadequate because of the severity of the health hazard presented by sulfites, because most sulfite uses were nonessential, with readily available substitutes, and because labeling would prove unworkable and unenforceable. In CSPI's view, only a ban could adequately protect consumers. This position, moreover, was consistent with the GRAS Select Committee's conclusion that "When benefits are related not to health, but to organoleptic, technologic, or economic considerations, substantial risk even to a relatively small subgroup of the population is not generally acceptable."[53]

Representatives of the Grocery Manufacturers Association (GMA) also spoke against labeling, but from a quite different perspective. GMA representative Sherwin Gardner, a former deputy and acting commissioner of FDA, emphasized that the class of sensitive individuals was small and that concern for them, however real, should not affect the safety status of sulfites for the general population. Gardner, too, sought support from an earlier SCOGS report, but the proposition he cited was that "special consumers require special considerations."[54] This statement implied, he said, that the GRAS status of sulfites should not be determined solely by the adverse reactions of a small minority. Gardner's recommendation was that the panel should simply reaffirm the 1976 SCOGS finding that sulfites present no risk to public health when used at current levels and according to current practices.[55]

The National Restaurant Association (NRA), an organization representing approximately 100,000 food outlets, adopted a position somewhere between those of CSPI and the GMA. The NRA spokesman strongly seconded CSPI's assessment that labeling was not a practicable policy approach and testified that sulfite use was already decreasing in the food service industry. To protect restaurateurs from the practices of their diverse suppliers, however, NRA believed FDA should impose a partial ban on sulfites, targeting those uses that present greatest risk to sensitive groups.[56] In NRA's view, the most dangerous applications of sulfiting agents were on salad ingredients and precut potatoes at both the retail and wholesale distributor levels.

These statements influenced the sulfite panel's final report in two significant respects. In the first place, the panel agreed with CSPI and NRA that "additional labeling requirements alone would not assure protection" against certain uses of sulfites and urged FDA to encourage the discontinuance of these uses through "appropriate use of the regulatory process."[57] Second, with regard to the health risks of sulfites, the panel felt unable simply to reaffirm the SCOGS report as urged by the GMA, although it was equally reluctant to make definite findings about the risk to sensitive individuals. The evidence before it, the panel felt, did not conclusively establish a cause-effect relationship between consumption of sulfite-containing foods and the onset of adverse reactions. However, the reported associations were sufficiently numerous and serious (including four reports of deaths of asthmatic individuals) to deserve serious attention.[58] Accordingly, the panel decided to address the question of risk separately for the general population and the class of sulfite-sensitive individuals. With respect to the former, the panel felt, as before, that there was no evidence of a hazard at current levels and conditions of use. But for the latter the panel deemed it necessary to add an explicit caveat:

> For the fraction of the public that is sulfite sensitive, there is evidence in the available information . . . that demonstrates or suggests reasonable grounds to suspect a hazard of unpredictable severity to such individuals when they are exposed to sulfiting agents in some foods at levels that are now current and in the manner now practiced.[59]

The conclusions of the ad hoc panel, together with recent medical reports, consumer complaints, and comments from the public on the 1982 proposed rule, prompted FDA to reconsider its policy regarding sulfites. In July 1986, FDA decided to revoke the GRAS status of sulfites for use on raw fruits and vegetables,[60] but to allow their continued use in prepared potatoes, dried fruits, shrimp, and wine. In a parallel ruling, FDA also required that processed foods containing more than 10 ppm of sulfiting agents should be labeled so as to disclose their sulfite content.[61] The final regulatory package thus represented a compromise between consumer and manufacturer interests, as well as a change from FDA's initial policy proposal.

Was it necessary for FDA to convene a scientific advisory panel in order to reach these results? Assessments vary, depending on the institutional affiliations of the evaluator. From FDA's perspective, the recourse to FASEB and the ad hoc panel was a justifiable move, given the need to evaluate the new medical information on sulfites and the agency's general practice of involving the scientific community in potentially controversial science policy issues. Satisfied by their success in producing an expert consensus, neither FDA nor the FASEB staff who convened the sulfite panel seriously questioned the need to seek advice. CSPI, however, characterized the process as an unnecessary diversion on a matter that the agency could easily have resolved on its own with the aid of in-house expertise.[62] In short, the sulfite case reinforced the perception of public interest groups that regulatory peer review is simply a ploy to delay or avoid politically unpalatable decisions.

The facts provide some support for CSPI's analysis. The sulfite panel carried out no scientific investigations, developed no new analytical methods, and came to no conclusions that went beyond the limits of FDA's technical competence. Moreover, one of the panel's most influential determinations—that labeling was not an adequate remedy for the most hazardous uses of sulfite—involved neither scientific nor medical judgment. It was a policy conclusion, based largely on an assessment of practicability that FDA was wholly capable of reaching without expert assistance. Finally, the panel's most important contribution to the dynamics of decisionmaking had little or nothing to do with its specialized expertise. The public hearing conducted by a seemingly neutral party had the salutary effect of defusing what could otherwise have developed into a showdown between FDA and consumer or food manufacturing interests. All parties were afforded an opportunity to influence the panel's determinations, thus gaining a sense of having participated meaningfully in the decisionmaking process.

Yet it would be misleading to suggest that the panel's scientific credentials were irrelevant to the legitimacy of the regulatory outcome. In the first place, the panel's expertise bolstered the credibility of the conclusion that sulfites posed a health threat deserving of attention, especially since the evidence on this issue was not conclusive. FDA's decision to limit the sulfite ban to raw fruits and vegetables also drew support from the panel's finding that these were the most problematic uses of the compounds. Although relatively little of the panel's public work was scientific in form or substance, it is unclear whether FDA could successfully have reached the same results without invoking the authority of impartial science. In any event, the absence of serious criticism or litigation following the partial ban suggests that the panel's hybrid review process assisted FDA in arriving at a stable policy consensus.

Color Additives: An Endless Scientific Frontier

The case of color additives presents an interesting contrast to sulfites, for here scientific advice arguably acted as an impediment to consensus-building. Color additives appeared on FDA's regulatory agenda in 1960, when the Color Additive Amendments to the FDCA provided that new color additives could be approved for marketing only if FDA determined with reasonable certainty that their intended uses posed no risk to human health.[63] The amendments included a version of the Delaney clause, prohibiting any color additive shown to be carcinogenic in man or animals from being approved for commercial use.[64] For additives already in use in 1960, the amendments established a system of provisional listing pending further safety studies. Approximately 200 color additives were originally involved in this scheme. Of these, 126 were permanently listed (that is, declared safe for use) following additional testing, 63 were eliminated as unsafe, and about 10 still remained on the provisional list more than twenty years after the passage of the amendments.[65] Toxicological evidence from animal studies indicated that the Delaney prohibition might apply to as many as six of the ten additives: D&C Red No. 19 (R–19), D&C Red No. 37 (R–37), D&C Orange No. 17 (O–17), D&C Red No. 9 (R–9), D&C Red No. 8 (R–8), and FD&C Red No. 3 (R–3). The fate of these six additives became a source of considerable embarrassment to FDA during the 1980s.

The Appeal to Peer Review

Recourse to peer review in the case of the six color additives marked the terminal stage of a long-drawn-out period of incremental decision-making. It was not until 1976 that FDA completed its initial review of existing color additives to determine whether they satisfied the safe-for-use requirements of the 1960 amendments. Even then, the agency admitted that scientific data on fifty-two compounds were too scanty to permit an adequate assessment of their safety.[66] Accordingly, FDA mandated additional testing on these compounds and, in 1981, created a set of staggered closing dates for some two dozen additives that still remained on the provisional list. This schedule corresponded to the expected completion dates of chronic toxicity studies then under way on the substances. In March 1984, Mark Novitch, then Acting Commissioner for Food and Drugs, recommended to Dr. Edward Brandt, the Assistant Secretary for Health, that provisional listing for the six compounds be permitted to expire three months later. The basis for this recommendation was FDA's determination that the additives caused cancer in laboratory animals, thus violating the Delaney clause.

At this point, the Cosmetic, Toiletry, and Fragrance Association (CTFA), a trade group, raised a number of objections to banning the six suspected carcinogens. CTFA's primary argument was policy-based. The association asserted that over the previous decade FDA had established as a policy matter that the Delaney clause did not have to be applied to substances which, even under the most conservative risk assessment, posed no additional risk of human cancer. According to CTFA, the principle that such *de minimis* risks need not be regulated was approved not only by the D.C. Circuit in *Monsanto v. Kennedy,*[67] but also in a series of agency actions concerning animal drug residues, food and color additive constituents, and a hair color.[68] CTFA claimed that its own risk assessments on the six additives in question showed that any risk they presented fell below the level of one in one million that FDA recognizes as *de minimis*.

A second factor cited by the trade association was that FASEB, under an independent contract with the National Cancer Institute, was then in the process of reviewing the carcinogenicity of two of the six compounds targeted for delisting by FDA. FASEB's interim report indicated that the animal carcinogenicity of the substances was well established, but the report expressed a number of reservations about using the animal data to draw conclusions about human risk.[69] For example, the report noted that the metabolic pathway by which the dyes produce tumors in rats made it "very difficult to extrapolate or estimate any possible human carcinogenic risk."[70] The report also called attention to the absence of any epidemiological studies linking the dyes to effects on human health.

In spite of FASEB's cautionary notes, FDA recognized that the finding that the additives caused cancer in animals was legally sufficient to require delisting under the established reading of the Delaney clause.

The agency maintained publicly and in correspondence with CTFA that the Delaney clause did not permit it to ignore colors known to be animal carcinogens simply because they presented very low risks, and that it had no intention of construing *Monsanto* as a broad *de minimis* exception to the Delaney principle.[71] Under questioning by a congressional committee, Dr. Brandt again confirmed that he supported the "rigid construction of the Delaney clause as it currently stands" and that he in no way proposed to alter FDA's or his own interpretation of that provision.[72]

But CTFA also argued that FDA had recently become aware of two new scientific issues relevant to risk assessment: the theory that substances may be differentially absorbed by test animals and humans ("selective penetration"), and the possibility that complex mixtures such as food or color additives might vary in their toxicity according to the precise composition prepared by each manufacturer. Both issues, in CTFA's view, had a potentially important bearing on determinations of the carcinogenicity of color additives, and the trade association argued that a ban was inappropriate until these issues were peer-reviewed by an institution such as FASEB.

In the judgments of both Dr. Brandt and Dr. Frank Young, the newly appointed FDA Commissioner, these combined questions of science and policy justified a further round of scientific review. Dr. Brandt determined, in particular, that CTFA had presented persuasive reasons for reconsidering the whole issue of risk assessment, and he manfully attempted, through a species of boundary work, to distance this issue from the question of reinterpreting the Delaney clause:

> But it seems to me that it's also of value to say, "Look, arguing de minimis makes no sense in this case because it's scientifically impossible to do a risk assessment." That's the question. FDA says it is. They say, "Don't argue de minimis, because in fact you can't do a risk assessment anyway." The CTFA makes the other argument, that that's not correct. Now, which is right?[73]

Brandt seemed to be saying that the real science policy issue before the agency was not "are the additives carcinogenic" (a foregone conclusion), but "is it possible to assess their risk." Pursuing this new line of reasoning, the agency concluded that it should seek additional expert advice on whether CTFA's approach to risk assessment was scientifically valid. Instead of approaching FASEB, however, the FDA Commissioner appointed an ad hoc Color Additive Scientific Review Panel "to evaluate: (1) the possibility of performing a scientifically valid carcinogenic risk assessment of the six colors for drug and cosmetic uses; and (2) whether the available information supports the risk analyses before the FDA."[74] The panel was not asked to reconsider whether the colors in question induced cancer in laboratory animals.

The color additives panel generally confirmed the validity of CTFA's approach to risk assessment. The risk estimates produced by both CTFA and the panel were low, reflecting the low levels of human exposure to these additives, especially in cases of external application.[75] To the extent that the assumptions used by the panel could be quantified, they produced estimates that were less than one order of magnitude lower than CTFA's. This relatively close concurrence led the panel to conclude that any possible overestimates and underestimates of risk made by CTFA on the whole canceled each other out. Putting the panel's work together with that of CTFA, FDA was thus in a position to conclude that the health risks of exposure to the six color additives would be negligible, at least if recent usage information proved to be reliable.

The panel's claim to credibility lay chiefly in its membership, which drew upon several Public Health Service agencies, though FDA and NCTR were most heavily represented.[76] Accordingly, its conclusions could be represented as a broad consensus of government scientists while remaining fundamentally in sympathy with FDA's views. Moreover, the panel took care to point out that its analysis was carried out against the backdrop of, and presumably in harmony with, other recent agency activities aimed at consensus-building on risk assessment.[77] Finally, the panel's report, explicitly laying out its risk-assessment assumptions, was published in a peer-reviewed journal, thus undergoing a round of formal professional scrutiny before entering the public domain.

Nevertheless, the media and the public remained skeptical of the reasons for the panel's involvement. A *New York Times* story reported that

> After 25 years of study and, by one count, 27 extensions of the regulatory deadline, the Food and Drug Administration has appointed a panel to review research indicating that six widely used food and drug dyes may cause cancer. Meanwhile, the food, drug and cosmetic industries, which contend that the dyes are safe, can continue to use them, the F.D.A. said.[78]

Accounts such as this did little to dispel the suspicion that FDA had co-opted the procedures of science for political purposes, and that the panel was appointed mainly because the DHHS Secretary had bowed to industry pressure not to regulate the six color additives.

A Judicial Rejoinder

If FDA was hoping that peer review would legitimate a major reinterpretation of the Delaney clause, its hopes were dashed by the D.C. Circuit. The agency decided in August 1986 to list two of the controversial dyes, Orange No. 17 and Red No. 19, in spite of the finding that these substances caused cancer in test animals. To justify this departure from prior practice under the Delaney clause, FDA asserted that quantitative risk assessment had shown the health risk from these substances to be *de minimis* and claimed that it had "inherent authority" to exempt such risks from regulation. In *Public Citizen v. Young*[79] the court was asked to consider whether this decision could be countenanced as lawful; somewhat reluctantly, it concluded that no such exception could be recognized.

The court's rationale had virtually nothing to do with the validity of FDA's scientific argument, deriving instead from a straightforward legalistic reading of the language, statutory context, and legislative history of the Delaney clause. A review of these factors convinced the court that Congress had intended to promulgate an absolute rule, one that could plausibly be read in only one way: "if the Secretary finds the additive to 'induce' cancer in animals, he must deny listing."[80] Since this finding had in fact been made with respect to both of the dyes in question, and was not contradicted by the color additives panel, the decision left FDA very little leeway to reach any other conclusion.

In an effort to forestall this literal interpretation of its mandate, FDA (with help from the Department of Justice) had made an intriguing attempt to redefine the concept of "inducing cancer" in animals by weaving quantitative risk assessments into the interpretation of bioassays. The agency suggested, in brief, that it was appropriate to take risk assessments into account when determining whether laboratory observations of carcinogenicity should be regarded as "inducing" cancer; a substance producing only minimal risk to humans should not be treated as a cancer-inducing agent within the meaning of the Delaney clause. The court, however, definitively rejected this argument, observing that "Congress did not intend the FDA to be able to take a finding that a substance causes only a trivial risk in humans and work back from that to a finding that the substance does not 'induce cancer in . . . animals.' "[81]

The *Public Citizen* decision undoubtedly came as a disappointment to those who believed that the problem of the Delaney clause could be resolved by fitting a scientific straitjacket over the irrational edges of the law. Yet a prescient observer might have predicted that FDA's attempt to recast the Delaney debate in purely scientific terms would simply invite the judiciary to retreat to its own boundaries and to respond with an insistence on the letter of the law. It was unrealistic to expect that unity of opinion between FDA and its technical advisers would be accepted by the court as sufficient grounds for overriding a settled legal interpretation. The D.C. Circuit in this case expressed no reservations about the technical validity of the risk assessments carried out by CTFA or FDA's ad hoc peer-review panel. It simply concluded that these assessments were irrelevant for legal purposes, and that legislative, not scientific, politics would have to provide the relief sought by the agency.

Advice and Decision

Given substantial discretion to structure its relations with scientific advisory committees, FDA has selected a variety of approaches that have, superficially, little in common. In the area of drug evaluation, the agency has constructed a reasonably systematic process, using an array of standing committees for routine assistance with NDAs and INDs. By contrast, the advisory process with respect to food and color additives remains much less standardized. FDA's relationship with FASEB can be mobilized as needed when the agency is confronted by new data on the hazards of food additives. But FDA is under no compulsion to consult FASEB and may even find it expedient to use safer sources of advice, such as NCTR, when dealing with particularly controversial issues. Practices of appointing committee members vary, ranging from relatively close control by the agency over the influential Cardio-Renal Committee to far less formal supervision of committees selected by FASEB. The timing of review, the procedures for committee meetings, and the issues addressed by advisers also differ from one FDA program to another.

What most of FDA's advisory processes have in common, however, is a blurring of the boundaries between science and policy. The agency freely permits its expert advisers to receive evidence from nonscientists as well as scientists and to deliberate on such topics as the appropriate cost of producing more information or the acceptable risk level for hypersusceptible consumers. Advisory proceedings, moreover, are structured in ways that leave the line of demarcation between FDA's decisionmaking authority and that of its scientific advisers ill defined. For example, in most drug-approval proceedings, advisory committees review the scientific literature and make recommendations before FDA develops a regulatory proposal of its own. Accordingly, when the agency makes a formal decision, its apparent effect is to ratify a consensus already arrived at by experts. In other cases, advisory committees not only conduct the review and evaluation of the scientific literature but also prepare the equivalent of a preliminary position statement. This was the role played by the Cardio-Renal Committee in its review of clinical testing guidelines for antiarrhythmic drugs and by the ad hoc sulfite panel in its assessment of the health risks of sulfiting agents.

At its best, this approach to seeking advice works well at reducing conflicts over the interpretation of regulatory science. A nonadversarial advisory process offers an ideal forum for negotiating among different scientific and political viewpoints. Discussion proceeds without polarization and without falling into the traps of experimenters' regress or endless scientific deconstruction. The ambiguity of the boundary between science and policy is also strategically useful to FDA, permitting the agency to harness the authority of science in support of its own policy preferences. As the propranolol case intriguingly demonstrates, it is possible for FDA to engage in creative boundary-work to expand the extent of its authority and gracefully deviate from committee recommendations. The color additives case provides an interesting counterpoint. Here, FDA tried to draw an unusually definite boundary between the scientifically and legally "correct" readings of the Delaney clause, with the result that its own credibility suffered and the power to interpret the provision passed ultimately into the control of the courts.

What, if any, are the drawbacks to FDA's style of seeking science advice? Purists might object that the blending of science and policy in advisory proceedings cedes too much political control to the experts, undermining the agency's accountability to the public. Put differently, one could argue that FDA's methods of consultation accord too much deference to the construction of evidence and risk by scientific experts who are not fully sensitive to the agency's legal and political mandates. The cases considered in this chapter (and in Chapter 2) suggest, however, that FDA in fact retains many opportunities to avoid being captured by its advisory network, at least when the agency is led by a strong commissioner and supporting staff. Through selective use of committees, public review of advisory reports, and occasional overruling of recommendations, FDA maintains the freedom to construe science and to make policy in ways that may in the end serve the public better than an overly rigid commitment to separating science from policy.

9
Coping with New Knowledge

Accounts of regulatory controversies often create the impression that science can be frozen at particular moments in time: the time at which the agency completed a risk assessment, consulted an advisory committee, or announced a policy decision. The realities of regulatory science, however, are generally far more complicated. In the several years consumed by a complex rulemaking process, the state of knowledge undergoes constant redefinition, often as a result of purposeful scientific activity by parties to the proceeding. The dimension of change has to be factored into any comprehensive account of science policy, for it bears on key aspects of the relationship between knowledge and politics. To what extent is change in regulatory science a socially constructed "reality" that meets the needs of one or more interested political actors? What role do advisory committees play in certifying that change has occurred or that it is relevant to policy? If scientific change is a construct, can it ever be meaningfully regarded as a justification for policy change?

Regulatory proceedings that cut across political and administrative boundaries offer a fruitful avenue for exploring these questions. In such cases, agencies often engage in multiple rounds of consultation with the scientific community. Expert advice thus serves not as a one-shot technique for legitimating the regulator's use of science but as part of an ongoing process of social accommodation among the interests (including science) that are affected by a far-reaching regulatory initiative. This chapter examines the role of scientific advisers in two such proceedings: the fifteen-year effort by several federal agencies to develop principles for assessing carcinogenic risk, and the only slightly shorter multi-agency initiative to reduce exposure to formaldehyde.

As two of the most contentious regulatory controversies of recent years, both the guidelines for cancer risk assessment and the exposure standard for formaldehyde have been widely discussed in the policy literature. The object here is not to go over this entire territory in detail but to focus on the precise role of expert advice in evaluating claims concerning scientific change. In each case varied types of advisory committees played an indispensable role in redrawing the boundaries of accepted science and shifting the applicable scientific consensus. Through the advisory process, competing claims about what was known or needed to be known were gradually sorted out and new scientific assumptions and conclusions were eventually incorporated into regulatory decisions.

The Quest for Principled Risk Assessment

In September 1986 EPA issued its final Guidelines for Carcinogen Risk Assessment.[1] The action was the culmination of a process that had begun more than a decade earlier, a decade marked by bitter debates among the agency, the chemical industry, the scientific community, and, finally, the Office of Management and Budget. During much of this period, EPA's attempts to place carcinogen risk assessment on a principled footing was supplemented by the activities of other federal agencies. These collective efforts included several different interactions with scientists situated outside the government, employing procedures that ranged from informal notice and comment to public hearings and structured peer review. In describing this history, it is useful to distinguish three phases of policy development. In the first phase, the major actors were EPA and other regulatory agencies, and the Interagency Regulatory Liaison Group (IRLG), which attempted to coordinate the efforts of the individual agencies. The second phase was marked by a temporary cessation of these initiatives while the Office of Science and Technology Policy (OSTP) attempted to codify the state of scientific knowledge concerning carcinogenic risk assessment. In the third phase, the momentum again reverted to EPA, as the agency updated its existing guidelines to reflect new scientific views concerning cancer testing and carcinogenesis.

Phase One: Playing by Rules

Throughout the 1970s regulatory agencies were under intense pressure to impose strict controls on manmade substances suspected of causing cancer. Yet the techniques for identifying such chemicals and accurately determining their risk posed intractable problems. The use of animal data to predict risk to humans, in particular, was (and still is) a persistent source of controversy. Experts disagreed about the criteria for judging the validity of such tests and about their transferability to humans. Questions concerning the regulatory significance of bioassays repeatedly arose in the early years of environmental policy formulation. When does cancer induction in mice or rats provide genuine grounds for concern about human beings? How should decisionmakers deal with conflicting results from different test systems? What interpretation should be placed on the fact that responses in the exposed animals do not always correspond to epidemiologically observable effects in humans? When animals are exposed at high doses so as to detect relatively minor risks, what do the results imply for risks at much lower doses of exposure?

The initial response of EPA and other administrative agencies was to treat these issues as policy problems to be resolved through rulemaking. Between 1974 and 1976 EPA developed a series of so-called cancer principles to defend itself in lawsuits challenging its restrictive actions on DDT and other organochlorine pesticides.[2] EPA's legal staff prepared the first version of the principles from the reports of various expert committees and, eventually, from the technical record compiled by the agency in regulating the pesticides heptachlor and chlordane. When manufacturers objected to the use of these principles on scientific grounds, EPA turned to a well-known government scientist, Umberto Saffiotti of the National Cancer Institute, for review and modifications. Working with an EPA attorney, Saffiotti developed a more refined statement of the cancer principles, but Saffiotti's governmental affiliations cast doubt on his neutrality, and his recommendations failed to silence the growing controversy about the scientific validity of the principles. Both Velsicol, the principal manufacturer of heptachlor and chlordane, and USDA demanded that EPA's principles be peer-reviewed by a technically competent body independent of the regulatory agencies, such as a committee of the National Academy of Sciences.[3]

Perturbed by these objections, EPA decided to follow a more public and systematic approach to the development of an analytical framework for carcinogen regulation. A committee chaired by Dr. Roy Albert helped the agency to formulate a set of "interim" guidelines for assessing the risks and possible economic impacts of potential carcinogens.[4] These guidelines provided for the creation of a unit called the Carcinogen Assessment Group (CAG) to centralize EPA's in-house expertise on cancer. In response to a petition from the Environmental Defense Fund, EPA published a proposed policy for classifying and regulating airborne carcinogens.[5] The agency also became an active participant in IRLG's effort to develop a coherent, cross-agency approach to identifying carcinogens and assessing their risks.[6] Other agencies followed EPA's example. CPSC, for instance, issued its own guidelines for carcinogen risk assessment in 1978,[7] and OSHA, after a much more protracted rulemaking process, promulgated a "cancer policy" addressing carcinogens in the workplace in 1980.[8]

By and large, these efforts fell short of accomplishing their intended purpose, although they did lay the groundwork for later and more successful codifications. EPA's policy for airborne carcinogens, for example, was never published in final form, and the agency continued to operate with its "interim" risk-assessment guidelines until the mid-1980s. CPSC's guidelines were struck down by a reviewing court in 1978 for procedural irregularities.[9] OSHA avoided similar procedural errors by deciding to issue its cancer policy as a binding regulation developed through formal rulemaking. Shortly after the Reagan administration took office, however, OSHA stayed significant portions of the policy and requested additional public comments, thereby admitting that its entire initiative needed scientific and political renegotiation. IRLG's guidelines proved perhaps most resilient, at least in the sense that agencies informally continued to follow these principles in regulating suspected carcinogens. But the interagency approach initiated by IRLG was effectively repudiated by the Reagan administration, which dissolved both IRLG and its sister interagency unit, the Regulatory Council.

The antiregulatory temper of the early 1980s and a general retreat from environmental policy initiatives were in part to blame for these inconclusive outcomes. But there were other reasons why EPA, CPSC, and OSHA all failed to produce risk-assessment documents of more lasting significance. The most basic reason probably was the decision by each of these agencies to adhere to a strict version of the science policy paradigm in developing the principles and policies applicable to carcinogenic risk assessment. To the staff in each agency, it appeared that the uncertainties and evidentiary conflicts involved in identifying carcinogens and evaluating risk data could not be reconciled by science, but could be bridged only as a matter of administrative policy. Scientists, accordingly, were brought into the process of guideline development only as a source of information, not as adjunct decisionmakers with power to define the content of the cancer principles. OSHA's rulemaking proceeding illustrated the consequences of this approach most dramatically. Hearings held by that agency called forth testimony from some of the most distinguished experts on carcinogenesis, as well as from scientists affiliated with particular interest groups. As disagreements developed among these experts, OSHA's own staff undertook to find accommodations among the divergent viewpoints by preparing a composite regulatory document. The agency justified these regulations as administrative policy determinations not subject to further scientific review. But in pursuing this strategy, OSHA laid itself open to the charge that it had adopted an overly conservative, even unscientific, approach to resolving the controversies aired in the rulemaking process.[10]

The furor surrounding the OSHA cancer policy drove home the point that regulations were an inappropriate vehicle for expressing systematic principles of decisionmaking in an area of rapidly changing scientific knowledge. Charges of "freezing science" were leveled at OSHA even before the cancer policy was promulgated in final form;[11] although the agency tried to address these complaints by building more flexibility into the consideration of new evidence, the critics were not satisfied. The episode led federal agencies to abandon the strategy of framing their risk-assessment principles as legally binding regulations.

EPA's cancer principles were never cast in the form of regulations, but they nevertheless came under fire because of the domination of the risk-assessment process by a tightly knit group of CAG experts. CAG's hold on carcinogen risk assessment during the 1970s was likened by one agency official to "American influence at the end of World War II."[12] Between 1976 and 1983 CAG assessed the carcinogenicity of some 150 compounds, including such perennially problematic substances as arsenic, benzene, dioxin, coke oven emissions, and vinyl chloride.[13] But the scientific community's response to CAG's approach was by no means uniformly approving, and by the early 1980s CAG had been criticized by both the National Academy of Sciences and the SAB for excessive rigidity and conservatism in its risk-assessment methodology.

The primary insight that emerged from the public criticism of federal risk-assessment initiatives in the 1970s was that a new model would have to be found for involving the scientific community in the regulation of cancer-causing substances. A consensus developed that unsupervised administrative agencies were unlikely to evaluate the uncertainties of risk assessment in a sufficiently balanced manner to satisfy their scientific and industrial critics. EPA's experiences with CAG suggested that agency experts might prove unacceptably resistant to modifying their analytic methods in response to scientific change. The American Industrial Health Council (AIHC), a leading spokesman for the industrial viewpoint, began to lobby strenuously for a science panel to review all agency decisions relating to carcinogenic risk.[14] Although EPA continued to play a lead role in the further development of risk-assessment guidelines, it could no longer ignore the pressure to consult more closely with nongovernmental scientists.

Phase Two: The OSTP Initiative

The debate over carcinogen regulation entered a second phase with the 1983 National Research Council study of risk assessment, which made three noteworthy contributions to the policy process. First, and perhaps most important, the NRC report recommended that risk regulation should be conceived as two distinct analytical processes—"risk assessment" and "risk management"[15]—the former comprising primarily scientific decisionmaking, and the latter integrating scientific determinations with legal, economic, and political concerns. Second, the study concluded that risk assessment can never be completely separated from risk management, since bridging uncertainty in risk assessment always involves some elements of policy choice. Third, the study suggested that it would be unwise to create a single agency to carry out risk assessments for the federal government, displacing the decentralized efforts of individual regulatory agencies.

There was one area, however, where the NRC study conceded that greater centralization, accompanied by more expert analysis, could prove beneficial. In any risk assessment, there is a need for inferences based on incomplete data, but the inferences made by OSHA and other agencies were widely perceived as either inexpert or politically colored. The NRC panel concluded that conflict would be reduced and consistency fostered across regulatory agencies if a single, impartial expert board articulated guidelines for drawing the necessary inferences.[16] The Office of Science and Technology Policy (OSTP) decided to take up this challenge, although institutionally it was far removed from the body of independent scientists the NRC had in mind.[17]

The OSTP initiative grew out of inauspicious beginnings. By 1982 it was clear to all interested observers that the Reagan administration had its own ideas about where to go with carcinogen risk assessment, and that it would follow its agenda even at the cost of turning away from established agency practices. The first indication of the changes advocated by the administration became public when EPA in 1982 drafted guidelines for assessing the carcinogenic potential of water pollutants. As this document circulated it drew strong protests from environmental activists and scientists who had been involved in earlier stages of guideline development at EPA.[18] To many outside reviewers it appeared as though EPA was rejecting, without any fresh solicitation of scientific opinion, the consensus positions embodied in its own former guidelines and those drawn up by the IRLG. The move seemed all the more suspicious because on a number of critical points the new guidelines clearly catered to industry's interests.

Several prominent scientists, such as Norton Nelson of New York University and Bernard Weinstein of Columbia University, were incensed at what they viewed as a scientifically unfounded attempt to distinguish among potential carcinogens on the basis of their mechanisms of action. In brief, the new EPA strategy made a distinction between carcinogens that cause gene mutations and so-called epigenetic or nongenotoxic compounds.[19] EPA proposed to set exposure levels for the latter by applying a safety-factor analysis to no observable effect levels (NOELs). This approach had in the past been rejected for *all* carcinogens, since there was wide agreement that no safe thresholds of exposure could practicably be established for any of these substances. Congressional and media attention, coupled with the political scandals that were rapidly engulfing the Gorsuch administration, eventually led EPA to abandon these controversial departures from earlier initiatives.

When OSTP first assumed the lead on carcinogen risk assessment, it seemed to be operating with the same politically determined scientific agenda as the White House. Indeed, OSTP's initial statement of the scientific basis for assessing carcinogenic risk simply repeated many of the controversial positions adopted in EPA's 1982 proposal. A chapter written by John Todhunter, in particular, perpetuated the distinction between epigenetic and genotoxic agents that many in the scientific community found insupportable. The appearance of bias in the OSTP draft led Norton Nelson to comment before a Congressional subcommittee:

> I have no doubt that there are scientists in the federal establishment that can do quite as well in terms of the science as scientists outside the Federal establishment, but I do not believe that the credibility required in this important issue can be achieved by cancer policy formulated within a purely administrative framework of OSTP and the federal agencies.[20]

Soon thereafter the OSTP effort was reorganized under the leadership of Dr. Ronald Hart, director of the National Center for Toxicological Research. One of the operational changes introduced by Hart was a far more aggressive policy of scientific peer review.

The second external review to which the OSTP guidelines were subjected had much to do with establishing their ultimate credibility. Approximately 110 reviewers from all segments of the scientific community were consulted and 91 responded.[21] Table 9.1 summarizes the reviewers' reactions according to their professional affiliations, and Table 9.2 provides a percentage breakdown of positive and negative reviews for the two parts of the document.[22] The reviewers submitted

Table 9.1. Reviewer ratings of second draft of OSTP cancer risk guidelines.

Affiliation	Excellent	Very good	Good	Neutral	Negative	Total
Industry*	3	4	2	2	1	12*
Academia	9	18	4	3	1	35
National laboratories, international and national science organizations	4	3	—	—	—	7
Government	6	5	1	1	—	13
Others—consultants, contract labs, individuals	8	7	—	4	1	20
Total	30	37	7	10	3	87*

*Four reviewers with industrial affiliations gave different ratings for Part I and Part II. The responses were (Part I:Part II)—good:very good; neutral:very good; negative:very good; and negative:good.

Table 9.2. Percentage of positive and negative responses to OSTP cancer risk guidelines.

	Excellent + very good	Positive	Neutral	Negative
Whole document	77	85	11	3
Part II*	77	86	11	3
Part I*	74	82	12	5

*Combines results of different reviews for the two parts and reviews of the whole document.

several hundred pages of comments designed to fine-tune the guidelines, all of which the members of the task force sought to address in the next round of revision. The general rule guiding the preparation of the final draft was that "all statements had to be scientifically defensible and properly referenced." The resulting document was published in the *Federal Register* in May 1984 with a request for additional comments from the public.[23] Relatively few were received and these were largely supportive, suggesting that the published guidelines represented a real consensus about the current state of carcinogen risk assessment. Even critics of the first version conceded that wide review had led to a result about which scientists should have few qualms.[24] The final risk-assessment guidelines were published in the *Federal Register* in March 1985,[25] and republished as a journal article shortly thereafter.[26]

The upshot of these proceedings was to marshall the authority of science firmly in support of some basic principles of cancer risk assessment. OSTP, however, was careful to point out that these formulations were not ordinary science. Their hybrid character, for example, was openly acknowledged. Like the NRC risk-assessment study, the OSTP document admitted that "it is not possible to draw a sharp distinction in every instance between principles which are based solely on science and those which embody a choice based also on science policy." Finally, particular care was taken to explain that the principles were not a permanent scientific code: "It is clearly understood that new information and newly emerging concepts may modify some of these statements. Indeed, an unstated 'zeroth' principle is that regulatory judgments should embody an openness to advances in science and emerging scientific understanding."[27]

Phase Three: Consensus Versus Politics

Building the Consensus

The thorough housecleaning that followed Anne Gorsuch's departure in 1983 placed EPA's risk-assessment project under new intellectual and political leadership. William Ruckelshaus, Gorsuch's successor as EPA administrator, restored an atmosphere of openness, and the process of guideline development proceeded with greater attention to public and professional review. A revised draft of the guidelines for cancer risk assessment was published in November 1984; in contrast to earlier initiatives, this document was the product of consultation across the agency and not the work of the Carcinogen Assessment Group alone.[28] A package of draft guidelines (addressing carcinogenicity, mutagenicity, exposure, chemical mixtures, and development effects) was also submitted to the Science Advisory Board for formal peer review.

SAB was asked to evaluate both the overall adequacy of the scientific rationale for issuing the guidelines and EPA's resolution of certain technical issues relating to each guideline. To conduct the reviews, the Board appointed five groups of experts with specialized knowledge of the subject. These groups met with the agency and members of the public in March and April 1985 and transmitted their completed evaluations to the administrator in June 1985.

SAB's Carcinogenicity Guidelines Review Group rendered, overall, a very favorable assessment of its portion of the draft product.[29] Only two issues were identified as needing further elaboration, and in one of these cases the reviewers themselves provided clarifying language for the agency to use.[30] For the most part, reviewer comments were more narrowly aimed at improving the clarity, consistency, and comprehensiveness of the guidelines. The review group recommended, for instance, that the distinction between statements supported by scientific evidence and science policy be sharpened and that key terms such as "weight-of-the-evidence" be used consistently throughout the document.

Political Review: The Role of OMB

While the SAB report completed the scientific review of the risk-assessment guidelines, EPA still had to contend with a potentially more prejudicial review by the Office of Management and Budget. Beginning with the Nixon administration, OMB's involvement in the regulatory process had grown more intensive with each succeeding presidency. In particular, two Executive Orders (E.O.) issued by the Reagan administration had consolidated OMB's control over rule-making to an extraordinary degree. The earlier of these, E.O. 12291 of 1981, required all agencies to demonstrate to OMB's satisfaction that the social benefits of major regulatory actions exceeded their potential costs. This was followed in 1985 by E.O. 12498, which increased OMB's authority to supervise the prerulemaking stages of policy development. Agencies were required to prepare an annual overview of all regulatory actions and to have this regulatory program approved by OMB. Any subsequent deviation from the program would also require approval. Under the guise of reducing the economic cost of regulation and avoiding duplication and conflict across agencies, OMB thus acquired unprecedented power to ensure agency compliance with the administration's overall regulatory philosophy.

Within OMB, responsibility for implementing the two Executive Orders fell to the Office of Information and Regulatory Affairs (OIRA), a unit created by the Paperwork Reduction Act of 1980.[31] The Act stipulated that regulatory proposals and information-collection requests submitted to OIRA should be reviewed within ninety days or else be automatically approved. In practice, however, OIRA frequently exceeded this time limit,[32] and there were persistent complaints that the size, age, and inexperience of OIRA staff contributed to unnecessary regulatory delays.

OIRA's intervention in rulemaking also raised legal and constitutional questions. In a 1986 challenge by the Environmental Defense Fund, the D.C. District Court held that OMB could not use the regulatory review process to delay the promulgation of regulations past a statutory deadline.[33] House Energy and Commerce Committee Chairman John Dingell criticized OIRA review as a denial of due process in a 1986 hearing: "Time and time again, we have seen favored interests pleading their cause to OMB in secret meetings while other interested parties who played by the written rules were denied a similar opportunity to present their views or even to cross-examine the statements of those favored interests."[34] Partly in response to such concerns, OIRA agreed in 1985 to inform EPA on a trial basis of contacts with nongovernmental parties and to provide EPA with written records of those contacts.[35] Although OIRA Administrator Wendy Gramm consistently denied that E.O. 12291 was ever used in violation of statutory law, Democratic congressmen led by Representative Dingell succeeded in getting the House to vote to defund the unit in 1986. When the Senate refused to go along, the two Houses compromised on the position that OIRA funds could be used only for activities sanctioned by the Paperwork Reduction Act.

OIRA and the Construction of Risk

Although staffed primarily by economists and policy analysts, OIRA intruded into the scientific aspects of regulatory policy with little hesitation, involving itself between 1983 and 1986 in a series of science policy decisions under the jurisdictions of several federal agencies. Thus, in 1983 OIRA took the position that FDA need not ban Red Dye No. 19 under the Delaney clause since the risk was so low as to be nonexistent.[36] A year later, OMB's interference in OSHA's rulemaking on ethylene dioxide emerged as a legal issue when Public Citizen challenged OSHA's failure to issue a short-term exposure standard for that compound. The D.C. Circuit did not rule on the legality of OMB's position, but its dissatisfaction was evident in a decision to remand the case to OSHA for a better record on the effects of short-term exposures.[37] In 1986 OIRA took issue with an EPA proposal requiring businesses using new or modified steam boilers to report on sulfur dioxide emission tests. Gramm disapproved of the proposal partly on technical grounds, arguing that emission tests were inaccurate and would not produce any reductions in emissions.[38] In other instances, OIRA disputed the need for epidemiological studies relating to the health effects of environmental pollutants. In 1986 an EPA study of the effects of chemical residues in drinking water in a Wisconsin community was revised and resubmitted following OIRA objections to the study design.[39] Also in 1986, a Harvard School of Public Health study alleged that in reviewing proposed studies by the Centers for Disease Control, OIRA had systematically discriminated against projects in the areas of environmental and occupational health.[40] Finally, a scientist at the Goddard Institute for Space Studies charged in 1989 that OMB had altered his testimony to Congress to make his predictions regarding global climate change appear less certain.[41]

Given this record of intervention, it was hardly surprising that OIRA would actively concern itself with the issuance of EPA's risk-assessment guidelines. The cancer guidelines, as we have seen, had long been a topic of concern to chemical industry lobbyists, who enjoyed easy access to OIRA. Moreover, a questionnaire sent to various agencies in 1983 by a subcommittee of the House Committee on Science and Technology disclosed numerous instances of OMB involvement in risk-assessment policy.[42] EPA, for example, produced a memorandum from OMB criticizing EPA's standard-setting for toxic pollutants and recommending against the use of "worst case" estimates. Similarly, OSHA reported that OMB had questioned the risk estimates underlying a proposed standard for ethylene dibromide. FDA, consistently the most conservative of the regulatory agencies in its use of risk assessment, reported no similar interference. In addition, OMB was even alleged to have conducted its own risk assessment on acid rain and to have concluded that there was insufficient information to justify action on this front by EPA.

By 1986 there were reports in the media that EPA's process of developing risk-assessment guidelines was stalled because of opposition from OMB. These impressions were confirmed when Gramm made a speech before a chemical industry trade association stating that carcinogenic risks are often lower than estimated by various agencies and that she was considering asking her own staff to develop more reasonable guidelines.[43] These reports created a political furor that helped dislodge the proposed EPA guidelines from OMB's control and speeded up their official promulgation. Communications between EPA and OMB during this period suggest that EPA's prior efforts at peer review and consensus-building were also important in allowing the agency to win support for its position in this controversy.

In August 1986, Gramm wrote to EPA Administrator Lee Thomas expressing concern that the guidelines would not adequately inform decisionmakers of the uncertainties underlying risk assessments.[44] To improve the presentation of such information, she proposed a number of substantive changes. Four issues in the cancer guideline were identified as needing revision: selecting the most sensitive species and sex, counting benign tumors, choosing an extrapolation model, and making interspecies comparisons. In each case, Gramm objected to the presumptions that EPA proposed to adopt when information was lacking and suggested a reformulation that would either increase the decisionmaker's discretion to overlook the presumption or result in a numerically lower assessment of the risk. A comparison of EPA's proposal and OMB's suggested alternative on the issue of benign tumors is illustrative:

EPA: Benign tumors should generally be combined with malignant tumors for risk estimates unless the benign tumors are not considered to have the potential to progress to the associated malignancies of the same histogenic origin.

OMB: Benign tumors should be combined with malignant tumors for risk estimates to the extent that they are expected to develop into associated malignancies of the same histogenic origin.

Under OMB's formulation, benign tumors would less frequently be combined with malignant ones for purposes of risk assessment; this approach would have led in the aggregate to lower estimates of carcinogenic risk.

Lee Thomas defended EPA's formulations on the ground that they "were developed with full participation of scientists from government, academe, industry and public interest groups."[45] EPA's position, in other words, was presented as both scientifically and politically central, and OMB's, implicitly, as marginal. In addition, Thomas cited principles in the OSTP document and other scientific reports that militated against the specific wording changes recommended by OMB analysts. In the case of benign tumors, for example, Thomas noted that OMB's proposed formulation was too narrow, ruling out some scientifically defensible grounds for combining benign and malignant neoplasms. Again, EPA's wording was, in Thomas's view, more in keeping with the prevailing scientific consensus.

On the day that Thomas sent his reply to Gramm, EPA published the five risk-assessment guidelines in the *Federal Register,* an act that symbolically reaffirmed the agency's autonomy. In practical terms, the episode showed that political intervention was useless in an area where the agency clearly commanded the scientific high ground. In the end, OIRA failed to influence the policies embedded in the guidelines, although it unquestionably succeeded in delaying their adoption.

Formaldehyde: An Uncertain Carcinogen

EPA's Office of Toxic Substances has on its agenda a number of chemicals known as "buffalos" because of the intractable problems they present to policymakers. Formaldehyde, the twenty-sixth-largest-volume chemical in use in the United States, fully deserved this epithet. The compound was widely used in the manufacture of plywood and insulating materials, protective coatings, drugs, cosmetics, and textiles. Because of the chemical's pervasiveness, human exposure could occur by many pathways, including food, tobacco smoke, and occupational and environmental contact. A strong irritant, capable of producing allergic sensitization, formaldehyde had long been regulated in the workplace. In 1979, however, the flavor of the policy debate on formaldehyde suddenly changed when a study performed by the Chemical Industry Institute of Toxicology (CIIT) indicated that exposure to the substance produces nasal tumors in rats and hence presents a possible carcinogenic risk to people.

In response to the CIIT study, at least four regulatory agencies—EPA, CPSC, OSHA, and the Department of Housing and Urban Development—undertook to review their formaldehyde policies. The study itself was accepted as valid, but it reopened questions about cancer risk assessment as policymakers tried to assess its implications for standard-setting. The process of deciding what to do with this new piece of information led to repeated consultations between regulators and a variety of expert panels. A brief history of these initiatives provides insights into the pros and cons of different approaches to certifying knowledge change in regulatory science.

Modes of Analysis

The problems of interpretation presented by the CIIT study need not be elaborated here in great detail because they have been extensively described in a number of works dealing with federal policy on formaldehyde.[46] The key problem, for our purposes, was the shape of the dose-response curve for rats, which was steeply nonlinear.[47] This factor sufficiently distinguished formaldehyde from "typical" carcinogens to revive the controversies that are always latently present when risk estimates for human disease have to be extrapolated from animal data. Regulatory agencies were forced to consider, in particular, whether the nonlinearity of the rat data implied the existence of a species-specific mechanism of carcinogenesis. More generally, the shape of the curve raised questions about the best way to fit the observed carcinogenicity in animals to a mathematical model for predicting risk to humans.

In seeking to resolve these issues, federal agencies employed numerous mechanisms for gathering information and advice: expert advisory bodies, both governmental and nongovernmental, administrative rulemaking, consensus workshops, and formal peer review. For reasons discussed below, most of these procedures proved of limited efficacy; it took the better part of a decade to develop a viable regulatory proposal on formaldehyde, and the role of the scientific advisory system in facilitating these developments was at best ambiguous.

The Federal Panel

The first extensive review of formaldehyde's effects on human health following the CIIT study was carried out by a panel formed under the auspices of the National Toxicology Program at the request of CPSC and composed of scientists drawn from the major federal research institutes. The panel reviewed published and unpublished literature on all health effects of formaldehyde except acute toxicity and hypersensitivity, which had recently been considered in an NAS report. Panel members then considered a series of questions relating to the carcinogenicity, mutagenicity, and teratogenicity of the substance. In its final report, published in 1982,[48] the panel determined that there was definite evidence that formaldehyde is both mutagenic and carcinogenic. More significant for policy purposes was the panel's conclusion that "formaldehyde should be presumed to pose a carcinogenic risk to humans."[49]

The Federal Panel's findings were widely cited by environmentalists seeking stronger controls on formaldehyde, but their ultimate scientific and policy impact was more questionable. Since the panel did not carry out a quantitative risk assessment, the report provided no numerical estimate of how many excess cancers could be expected from human exposure to the compound. This was a major deficiency at a time when risk assessment had begun to play an increasingly important role in standard-setting. The panel also provided little reasoning to support its central conclusion that the substance should be presumed carcinogenic to humans. Indeed, the panel's conclusions with respect to human health risks seemed less a considered interpretation of the CIIT study than an endorsement of the general principle that in the absence of countervailing evidence all animal carcinogens should be presumed to present a risk to humans.[50]

Yet the aberrant shape of the dose-response curve in the CIIT study presented regulators with a large question mark that they could not afford to ignore. Was the anomalous effect on the rat's nasal tract species-specific? The formaldehyde industry was not prepared to let regulatory agencies evade this issue by deducing the risk to humans according to established principles of risk assessment. Arguments about the proper way to read the CIIT data became progressively more insistent and polarized as agencies moved closer to the point of decisionmaking. The Federal Panel's findings at the same time grew less and less relevant to the new locus of scientific and policy debate. The panel's report appears in retrospect to have been curiously misdirected, perhaps because it represented the thinking of only one sector of the scientific community (government science), one particularly ill adapted to taking account of changing scientific knowledge.

Failures of Rulemaking

The CIIT study provoked new rulemaking on formaldehyde by both CPSC and EPA. In their turn, each of these efforts went awry, although for different reasons and with different consequences.

The UFFI regulation. CPSC's jurisdiction over formaldehyde derived from its use as a home insulant in the form of urea-formaldehyde foam insulation (UFFI). By the time preliminary results from the CIIT study became public, CPSC had received numerous consumer complaints about acute health effects occasioned by the installation of UFFI in residences and mobile homes. The report that formaldehyde might also be carcinogenic gave CPSC an additional motive for targeting UFFI as a top regulatory priority and for expanding its risk analysis of the compound to include both acute and chronic health effects. Between 1979 and 1981 CPSC took several steps to build a technical record on formaldehyde: besides convening the Federal Panel, the agency held four public hearings to gather data, held a technical workshop, and commissioned measurements of formaldehyde levels in building materials.[51] CPSC also performed a quantitative risk assessment using the CIIT data and a computer program called "Global 79," which yielded an upper limit of 1.8 cancers for every 10,000 homes insulated with UFFI. Armed with these findings, CPSC voted in February 1982 to ban UFFI in homes and schools.[52]

The formaldehyde industry immediately challenged this proposal, choosing to litigate in the Fifth Circuit Court of Appeals, where decisionmaking had tended to favor business interests over regulatory agencies. True to industry's expectations, the court overruled CPSC on both procedural and substantive grounds.[53] Judge Clark faulted the agency's cancer risk assessment for failing to take any account of the negative epidemiological studies on formaldehyde and for relying exclusively on the CIIT bioassay. Displaying strikingly little understanding of the basic principles of animal testing, the court concluded that "it is not good science to rely on a single experiment, particularly one involving only 240 subjects, to make precise estimates of cancer risk."[54]

Gulf South has rightly been criticized both as an overstepping of the boundary between judicial and administrative decisionmaking[55] and as an example of scientifically flawed judicial reasoning.[56] For purposes of this discussion, however, the most significant feature of the case was that it signaled a breakdown in the deference shown by courts to expert agencies under the science policy paradigm. The Fifth Circuit appeared to be sending a message that in an area as clouded with uncertainty as cancer risk assessment scientific claims would have to be backed up by something more than the regulator's alleged expertise in order to command judicial respect.

The TSCA §4(f) decision. At EPA, formaldehyde was caught up from the start in the broader political controversy over risk assessment described above. By early 1981, the Office of Toxic Substances (OTS) had identified formaldehyde as a candidate for priority attention under Section 4(f) of TSCA and was in the process of drafting a notice so designating the compound. Under instructions from Administrator Gorsuch, however, EPA Deputy Administrator John Hernandez scheduled several meetings with representatives of the Formaldehyde Institute and the Chemical Manufacturers Association, which culminated in what the agency termed a "science court" proceeding. EPA officials insisted that these hearings were conducted solely to elucidate the scientific issues concerning formaldehyde, but others perceived them as an inexcusable departure from the norms of open rulemaking. The meetings were not publicly announced and were attended only by industry representatives, EPA staff, and a handful of invited scientists.[57] Further, no transcripts or records were made, so that it proved impossible to determine what had been discussed at the meetings and what influence, if any, they had had on subsequent decisionmaking.

The appearance of bias intensified when OTS head John Todhunter sent a memorandum to Gorsuch with his own analysis of the formaldehyde data, concluding that the evidence was insufficient to trigger review under TSCA Section 4(f). Todhunter's analysis not only deviated in notable respects from carcinogen risk analyses previously performed at EPA but, worse still, rested at crucial points on faulty or unsupported assumptions.[58] For example, Todhunter concluded that the negative epidemiological data on formaldehyde were entitled to greater weight than the CIIT bioassay, but he did not discuss the methodological deficiencies of these studies or the impossibility of drawing statistically useful conclusions from them. Similarly, Todhunter's conviction that the carcinogenic response in rats could be explained through a species-specific mechanism had no empirical support, and his assumptions concerning probable levels of human exposure likewise rested on questionable foundations.

The memorandum generated an unprecedented outcry, with academic analysts, public interest groups, the scientific community, and Congress all demanding an explanation for Todhunter's deviation from his agency's administrative procedures and risk-assessment policies.[59] The controversy over formaldehyde, however, was soon swallowed up in the larger controversy about political interest-peddling at EPA. Todhunter along with most other senior officials left the agency in 1983. Their departure paved the way for the new administrator, William Ruckelshaus, to try and rebuild the agency's shattered credibility. Not surprisingly, one of EPA's early actions was to reconsider the regulatory dossier on formaldehyde.

The Impact of Peer Review

The last important phase in the regulatory analysis of formaldehyde extended from EPA's decision to reopen the proceedings under TSCA Section 4(f) (formally commenced in May 1984) and the release of a risk assessment of the compound's health effects in April 1987. During this time, both the scientific base for regulation and EPA's policy of securing external scientific advice underwent major modifications. As the state of knowledge changed, peer review acquired increasing importance as a mechanism for sifting through conflicting claims. Three further episodes in the formaldehyde story illustrate the consequences of this development: the 1983 "consensus workshop," the completion of the NCI epidemiological study in March 1986, and, finally, SAB's review of EPA's formaldehyde risk assessment.

An Ineffectual Consensus

By 1983, EPA recognized that the scientific issues relating to formaldehyde had to be debated in a less adversarial and more technically credible environment than the rulemaking process. The search for alternatives led first to a so-called consensus workshop organized at OSTP's request by EPA and the National Center for Toxicological Research. To ensure a balanced review of the scientific issues, the sponsoring agencies invited about sixty U.S. and European experts from academia, government, industry, and public interest groups. Divided into eight issue-oriented panels, these experts were asked to comment on questions put to them by the workshop's executive panel. These ranged from requests to evaluate the strengths, weaknesses, and implications of existing data to questions about gaps in the available information and the need for further studies. In keeping with the theme of consensus, the chair of each panel was instructed to guide the discussion so as to foster agreement among the panelists.

Effective as it was in reaffirming the federal government's commitment to open policymaking, the workshop failed to resolve the key technical disagreements that had hampered earlier initiatives to regulate formaldehyde. Two of the most important issues from the standpoint of regulatory action were the interpretation of the epidemiological data and the choice of risk-assessment models for extrapolating from the CIIT study to humans. With regard to both, the panelists' primary concern appeared to be not to overinterpret the evidence. Indeed, one study of the formaldehyde decisionmaking process concluded that "there was substantial scientific disagreement about the carcinogenic properties of formaldehyde and the Consensus Workshop did not even clarify—let alone resolve—these differences."[60] Thus, the epidemiology panel's overall conclusion was that "it is not possible from the available epidemiological studies to exclude the possibility that formaldehyde is a human carcinogen."[61] The risk-estimation panel, for its part, concluded that the epidemiological studies were not sufficient to permit quantitative risk-modeling. While the panel agreed that a nonlinear model was needed to fit the data from the CIIT study, it failed to shed much light on considerations that should guide the agencies in selecting an appropriate model.[62] It is hardly surprising that such equivocal conclusions exercised little influence on policy.

Additional Epidemiology

During the seven years of uncertainty between the CIIT study and EPA's final risk assessment of formaldehyde, federal decisionmakers appeared never to lose hope that science[63] might eventually provide an unambiguous resolution to their dilemmas. The consensus workshop was built—mistakenly, as it happens—on the premise that agreement on the most controversial technical issues was possible and would provide the necessary scientific support for policy. The $1 million epidemiological study of formaldehyde undertaken by NCI in 1981 represented yet another attempt to let science lead the policy process, this time by providing a "definitive" answer as to whether formaldehyde presents a cancer risk to humans. But in spite of many safeguards, including formal peer review, the NCI study became embroiled in controversy and failed to yield the clear answer that its proponents had hoped to produce.

The NCI cohort mortality study was quite simply the largest investigation ever undertaken of the formaldehyde and cancer link in humans. It followed some 26,000 workers from 10 factories and attempted to establish the cause of death for 3,268 members of the group who had died by the study's end. The investigators found no overall increase in cancer mortality nor statistically significant increases in lung or upper respiratory tract cancers. They concluded, accordingly, that there is "little evidence" that formaldehyde causes cancer among exposed workers. This reassuring news was immediately and widely reported in the media, receiving front-page coverage in both the *New York Times* and the *Washington Post*.[64]

NCI had taken many precautions to ensure the scientific credibility of the study's findings. The work had its roots in NCI's earlier epidemiological research on embalmers exposed to formaldehyde. A feasibility study was carried out in order to develop a valid protocol for the cohort mortality study. Interested regulatory agencies, industry and union representatives, academic scientists, and members of the Federal Panel on Formaldehyde all had an opportunity to review the protocol, which was revised several times before the inception of the study.[65] NCI appointed an advisory panel of six academic scientists, who also reviewed the protocol and approved the study from a technical and methodological viewpoint. Nevertheless, when NCI released its final report, five members of the panel wrote to Aaron Blair, the principal NCI investigator, disagreeing with the study's overall interpretation:

> We believe that this study does not resolve the issue of whether formaldehyde is a human carcinogen under the conditions which existed for the exposed members of the cohort. The finding of a significant increase in the risk for lung cancer (for exposed hourly workers who were first exposed more than twenty years before their

> deaths) makes us particularly hesitant to characterize this as a study with little or no evidence for carcinogenicity.[66]

Labor unions, too, were dissatisfied with the study results, and pressure from them, coupled with the advisory panel's critical response, led to a congressional investigation into NCI's administration of the formaldehyde project.

This close critical scrutiny produced a typical case of experimenters' regress, as NCI's credibility was challenged on both political and scientific grounds. Detractors charged that the study design was flawed and the reporting of results was biased because NCI had collaborated too closely with industry in carrying out the project. The Formaldehyde Institute (FI), the trade association for the formaldehyde industry, funded Phase I of the study and collaborated in its design under a memorandum of agreement with NCI. The study's nine coauthors, moreover, included two scientists from Dupont and one from Monsanto, both FI member companies, besides two NCI members and four outside consultants. Even the advisory panel, critics observed, had been appointed only after objections were raised to the potential for bias in the original NCI-FI study design.[67]

NCI and industry representatives denied the charges of bias,[68] but the substantive questions about the interpretation of the study proved difficult to ignore. A key point was the issue of excess deaths from lung cancer noted by the advisory panel. While overall lung cancer deaths were not higher than expected, there was a statistically significant (32 percent) increase in such deaths among workers who had been exposed for twenty years or longer. This was a point that many viewed as inconsistent with NCI's finding of "little evidence" of a cancer risk. Another important point related to the study's "statistical power"—that is, the capacity of a study of this size to detect increases in mortality. Analysts of the NCI study observed that the risk from the compound may have been "too small to be detected even by an epidemiological investigation of 26,000 workers."[69] With respect to apparel workers, for example, EPA had estimated that, for every 100,000 workers, two more will die as a result of occupational exposure than would have otherwise:

> Clearly then even if every one of the 26,000 workers were followed until his death, the additional one-half death from cancer would not be detectable. A study of 26,000 workers is not, however, a study of

> 26,000 deaths: only 3,268 had died by the study's end. Among these 3,268, then, one might expect—but of course could not observe—only an additional 0.1 cancer death.

Because of these weaknesses, the NCI study, despite its methodological care, size, and cost, failed to generate a real consensus about the human cancer risk from formaldehyde.

The SAB Review

EPA's Office of Pesticides and Toxic Substances (OPTS), in the meantime, had completed its first fully quantified assessment of the health risks of formaldehyde, and in June 1985 this draft document, entitled "Preliminary Assessment of Health Risks to Garment Workers and Certain Home Residents from Exposures to Formaldehyde," was submitted for review to SAB's Environmental Health Committee. OPTS sought answers to seven specific questions about the quality and scientific validity of its analysis (see Table 9.3); more generally, it wanted reassurance that the document provided a "balanced evaluation which lays out all the information to inform the Agency's decisionmaking."[70] Perhaps the most crucial task for the committee was to

Table 9.3. Questions for SAB review of formaldehyde risk assessment.

Question 1: Does the risk assessment appropriately utilize rat nasal squamous cell carcinomas and polypoid adenomas for hazard identification? Have the uncertainties in the tumor responses been adequately characterized?

Question 2: Has the risk assessment conveyed to the reader the inherent variability in the risk extrapolated from the rat carcinoma data?

Question 3: Has OPTS adequately extracted from the existing epidemiology studies the useful qualitative and quantitative information relevant to the assessment of formaldehyde carcinogenicity?

Question 4: Has OPTS handled the area of nasal "physiological" barriers in an even-handed manner?

Question 5: How can OPTS realistically begin to incorporate relevant kinetic information into quantitative cancer risk assessments?

Question 6: Do ranges of "margin-of-safety" for non-carcinogenic end-points adequately express the degree of concern (risk) for potential human exposures?

Question 7: What guidance can be given concerning the evaluation of risks from carcinogenic and non-carcinogenic endpoints following expected human exposures to formaldehyde, given that it is a naturally occurring substance in the human body?

validate EPA's treatment of two areas of new knowledge: epidemiological data on formaldehyde carcinogenicity and pharmacokinetic data potentially bearing on the quantitative assessment of risk. A third issue on which the committee substantively criticized OPTS's approach was the treatment of data on benign tumors.[71]

With respect to epidemiology, the SAB panelists noted that the picture had changed substantially since the CPSC's evaluation of formaldehyde. The completion of three or four major studies, including the one carried out by NCI, suggested to OPTS as well as the SAB reviewers that the human data on carcinogenicity might now be regarded as "limited" rather than "inadequate," as determined by the International Agency for Research on Cancer (IARC) in 1982. Because of the agency's confused presentation of the data, however, the SAB committee felt unable to assess its analysis adequately. The reviewers recommended that OPTS reorganize the section dealing with the epidemiological studies and make a more aggressive effort to derive quantitative implications from the data. Until such a reanalysis was carried out, the committee noted, it would not be able to provide additional guidance;[72] this statement implied that the committee expected to look at the document again after OPTS complied with its recommendations, an expectation that was not in fact fulfilled.

The committee also expressed some dissatisfaction with OPTS's analysis of the two types of tumors observed in the CIIT study. The squamous cell carcinomas were considered the more significant, and the committee felt that OPTS had not emphasized these sufficiently. Further, the reviewers appeared to think that the agency had not considered the possible relationships between the (benign) polypoid adenomas and the squamous cell carcinomas with sufficient analytical clarity. The committee therefore responded to OPTS's first question basically in the negative. Its most serious reservations, however, related to the question about pharmacokinetics. Here, the committee concluded that the OPTS risk assessment would not be scientifically adequate without a fuller consideration of the newly available research results.

The Pharmacokinetics Panel

Between the publication of the CIIT formaldehyde bioassays and the completion of OPTS's draft risk assessment, cancer researchers devoted significant effort to understanding the biological processes by which carcinogens interact with DNA. By undertaking such "pharmacokinetic" modeling, scientists hoped to elucidate the fundamental processes of carcinogenesis and, in the words of one author, "to break open the black box between external exposure levels and the ultimate production of tumors."[73] One avenue of research that seemed particularly promising was to look at specific biological "markers," such as the formation of DNA adducts, in order to get a better measure of the dose delivered to the target tissues of test animals. Such findings were especially important to industry in the case of substances like formaldehyde that displayed sharply nonlinear dose-response relationships. Pharmacokinetic research promised to clarify processes taking place at high doses that might be irrelevant to the degree of carcinogenic response at lower doses. Substantial work on DNA adduct formation involving formaldehyde was accordingly undertaken by a group of CIIT scientists in the early 1980s. OPTS could have attempted to incorporate this work into its risk-assessment model, making necessary modifications in EPA's existing principles. Instead, OPTS chose to disregard the CIIT research as not yet reliable enough for use in this context.

The SAB reviewers, however, were not persuaded by OPTS's unsupported judgment that it was premature to consider the pharmacokinetic data published by CIIT.[74] The committee recommended that a special workshop be convened to consider the issue in greater detail. The Office of Toxic Substances accordingly contracted with Life Systems, Inc., an independent consulting firm, to bring together a team of experts in metabolism, DNA adducts, and statistics to consider whether the available pharmacokinetics studies did or did not provide information suitable for use in assessing the health risks of formaldehyde.

Following a two-day meeting, which included informal discussion with CIIT researchers, the expert panel concluded that it would be desirable to have an effective measure of the dose delivered to the nasal epithelium of the rat and that the CIIT study marked an "important first step" toward making such measurements. Nevertheless, the experts agreed that the study was not sufficiently validated to provide a suitable basis for quantitative risk assessment.[75] In a strongly worded letter to OPTS, the CIIT scientists took issue with these conclusions, charging, among other things, that the panel had not inquired in sufficient detail into supporting work that had already been undertaken at CIIT.[76] The pharmacokinetics committee's conclusions were attributed by others to scientific conservatism in the face of new evidence and to the luck of the reviewer draw (see Chapter 4). Since neither EPA nor its contractor explained the basis for selecting this particular group of experts, it is difficult to evaluate the validity of these explanations. Whatever its professional biases may have been, however, the panel fulfilled the function for which it was convened: to provide a nonagency (hence presumably "independent") opinion on the certifiability of a new, and controversial, scientific claim related to risk assessment.

EPA's Response

The final risk assessment issued by EPA in April 1987[77] made numerous placatory gestures toward the SAB Environmental Health Committee. For example, a lengthy appendix providing individual summaries of the epidemiological studies reviewed by the agency clearly responded to the peer reviewers' concerns about the lack of clarity in OPTS's earlier treatment of this information. Another appendix described the work of the pharmacokinetics expert panel and its conclusions with respect to the CIIT studies. The risk-assessment document also discussed possible relationships between the benign and malignant nasal tumors observed in various formaldehyde bioassays and explained the basis for the agency's decision not to combine these different endpoints for purposes of statistical analysis.

But while the style and presentation of the final risk assessment changed in response to the SAB review, the agency's substantive conclusions remained virtually unchanged. As in the draft assessment, OPTS held to the view that the epidemiological data provided "limited" evidence of a risk to humans and that formaldehyde was a "probable human carcinogen (Group B1)." On the significant question of how to use the new pharmacokinetic data, the agency did not seriously modify its initial position, although the final conclusions were perhaps stated in a more conciliatory tone than before the SAB review:

> As with many emerging areas of investigation there are bound to be disagreements among scientists. This is one of those cases. Additional work underway at CIIT using primates may resolve areas of disagreement. Until the issues raised concerning the Casanova-Schmitz study are resolved, this study will not be used as a basis for an alternate measure of HCHO exposure.[78]

In other words, the agency did not see any need to deviate from the linearized multistage model of risk assessment on the basis of the new mechanistic information generated by CIIT.

Conclusion

The histories of the guidelines for cancer risk assessment and federal policymaking on formaldehyde illustrate the difficulties faced by agencies in keeping abreast of changes in regulatory science. Perhaps the clearest lesson to be drawn from the two cases is that the external scientific community's participation is essential to the certification of new knowledge and its exclusion from or incorporation into policy. EPA, for example, was ultimately able to prevail in its struggle with OMB over the revised guidelines for cancer risk assessment largely because it was backed by the impressive authority of the SAB and by OSTP's peer-reviewed cancer principles.[79] Similarly, EPA's decision not to use pharmacokinetic information in the formaldehyde risk assessment gained acceptance only when this action was certified as legitimate by a group of nonagency scientists. Moreover, the agencies proved unable to achieve these results until they supplemented the normal administrative process with practices derived from scientific decisionmaking. The adversarial rulemaking approach, in which scientists functioned as just another interest group, failed in both cases to build the kind of technical support the agencies needed. It was only when scientists were consulted through the channels of consensus workshops and regulatory peer review that their opinions began coalescing in ways that were supportive of policy.

Both cases also illustrate a development long awaited by critics of the regulatory process: a shifting of the procedure for closing scientific debate from judicial review to peer review. This was perhaps most apparent in the formaldehyde case, where the SAB exerted a kind of control that courts are arguably barred from exerting pursuant to *Vermont Yankee*. Specifically, SAB recommended that EPA adopt an additional procedure—an expert workshop—to ventilate the issues relating to the pharmacokinetics of formaldehyde. EPA complied, thus committing both more time and more money to the risk-assessment process than originally planned. To be sure, judicial intervention in this case might have led to similar results. *Vermont Yankee* has generally been interpreted to bar courts from imposing new procedural obligations on regulatory agencies, but administrative decisions may still be remanded for deficiencies in the underlying record, such as an agency's failure to respond to significant scientific claims by one of the parties. CIIT's pharmacokinetics data were so relevant to the risk assessment of formaldehyde that even a post–*Vermont Yankee* reviewing court might have, at industry's request, ordered OPTS to seek independent advice on this issue.

While the review processes in both cases were useful in cementing a new scientific consensus, their impact on policy was much more ambiguous. It could be argued, for instance, that EPA's garnering of expert support for the risk-assessment guidelines exacted a very significant price. The final guidelines were so flexible that it was unclear whether they would genuinely speed regulatory decisions. Alhough their original purpose was to establish a principled, generic approach to risk assessment, the guidelines that EPA eventually adopted seemed to call for the very case-by-case consideration of difficult science policy issues that the agencies had once been at such pains to avoid.

OPTS's final risk assessment for formaldehyde likewise escaped serious challenge, but by the time it was completed the torch of regulatory initiative arguably had passed out of EPA's control. In March 1986, EPA officially recognized that lead responsibility for protecting workers in the textile industry belonged with OSHA and referred this issue to its sister agency under Section 9 of TSCA.[80] This left the relatively less significant issues relating to exposures in pressed wood manufacture under EPA's jurisdiction. OSHA finally promulgated a 1 ppm standard for formaldehyde in November 1987, but this action came years after several European countries had lowered their workplace exposure limits for the substance with considerably less acrimony and commitment of resources.[81] Skeptics could well ask, therefore, whether the cessation of scientific controversy in the United States was a product of new knowledge, certified by improved review procedures, or a reflection of the formaldehyde industry's gradual admission that lower standards were technically achievable.

10
Technocracy Revisited

Despite the trend to wider use of advisory committees, regulatory science continues to be debated and interpreted within the structural constraints of an administrative process that still is heavily dominated by politics and law. Politically appointed officials, accountable not to science but to Congress, the White House, and the courts, control key decision points in the advisory process, from making initial committee appointments to deciding whether or not to follow advice. While regulatory agencies concede that legal and political considerations should not influence the design or conduct of policy-relevant research, as happened at Love Canal, most such work continues to be carried out under the auspices of government or industry, both of which have stakes in constructing scientific knowledge to fit particular policy ends. For committed advocates of the technocratic model of science policymaking, this picture represents at best a partial victory. The problems of credibility and conflict that continue to beset regulatory science require, in their view, more radical responses than sporadic, and largely discretionary, consultation between agencies and advisory committees.

Assuring the independence of research is one of the cardinal objectives of the technocratic vision, as well as one of the least controversial. If regulators and political interest groups wielded less influence on the production of policy-relevant science, so the technocrats would argue, the results would be more readily acceptable to a wider range of actors. Although often coupled with a naively positivistic view of scientific knowledge, this argument makes a certain amount of sense from a social constructivist viewpoint as well. Knowledge generated by parties without a clear stake in its application might indeed be more resistant to the deconstructive pressures of U.S. regulation. This at any rate is the hypothesis underlying the distinctive research program created by the Cambridge-based Health Effects Institute (HEI), an organization funded by EPA and the automobile industry. HEI's efforts to build a politically neutral and scientifically impeccable knowledge base for use in regulating air pollution deserve consideration in any discussion of science policy reform.

The technocratic approach to science policy becomes considerably more controversial when it seeks to give scientists a larger share of decisionmaking authority on issues with a clearly discernible policy component. Yet U.S. scientific advisers have played such a role, albeit on a limited scale, in numerous areas of federal policymaking. FDA, in particular, has on several occasions delegated the responsibility for conducting risk assessments, and even recommending policy, to scientific societies such as NAS and FASEB. Another approach, again drawn from FDA's decisionmaking repertory, is to substitute expert panels for administrative law judges, creating, in effect, miniature science courts for adjudicating the technical claims of regulatory adversaries. FDA's Public Board of Inquiry (PBOI), a procedure first used to evaluate the safety of the artificial sweetener aspartame, illustrates the possibilities and limits of this approach.

Initiatives such as these force us to reconsider how deeply the U.S. regulatory process really is committed to the concept of separating science and policy. These novel arrangements certainly indicate that the system is capable of tolerating relationships between scientists and policymakers that diverge significantly from the canonical science policy paradigm toward the technocratic model. The remainder of this chapter seeks to ascertain to what extent each of these arrangements is a viable alternative to the services provided by conventional scientific advisory committees.

A Public-Private Partnership for Science

HEI was the brainchild of Charles Powers, a former ethics professor and business executive, who became convinced of the need to reduce scientific conflict in the promulgation of clean air standards.[1] Powers embraced two simple—but, in the American context, revolutionary—prescriptions for enhancing the legitimacy of regulatory science. First, he concluded that the level of controversy surrounding science policy could be decreased if research funding decisions were taken out of the adversarial dynamics of the regulatory process. Second, he believed that a research-sponsoring institution acting independently of government and industry could produce science of such high caliber that none of the parties to regulation would question its validity.

The concept that Powers began developing in 1978 was an institute that would be equally funded by government and industry and would serve as a conduit for reviewing proposals and awarding grants. The organization's credibility would derive from its ability to formulate and fund first-rate research, with backing from a distinguished board of directors (HEI's board originally included such public luminaries as former Watergate Special Prosecutor Archibald Cox and Stanford President Donald Kennedy) and standing advisory committees of noted scientists. Powers's objective, in short, was to establish an institution whose prestige would be sufficient to guarantee its place as the nation's premier provider of information about the health effects of air pollutants.

When HEI was formed in 1980, its funding, amounting to about $6 million a year, was obtained half from EPA and half from dues paid by some two dozen companies manufacturing or selling automobiles and automotive engines in the United States.[2] This partnership brought together two of the actors with the largest stakes in the field of air pollution control and gave them a common interest in the success of HEI's research program. But in order to establish its credibility with the public, the Institute also had to demonstrate its independence from its financial sponsors. HEI sought to do this with the aid of two scientific committees, one to supervise the awarding of research grants and one to review research results. This bifurcated system for obtaining peer review proved the key to HEI's success in persuading outsiders of its objectivity.[3]

HEI's Peer-Review Process

HEI's Health Research Committee is responsible for defining the organization's research goals and priorities and for awarding contracts. Though it considers recommendations from HEI's cosponsors, the committee is not bound to honor their requests. In fact, in an early assertion of independence, the Board of Directors and the Research Committee walked out of a meeting with EPA and industry rather than submit their research agenda to review by the sponsors. In 1982 the Research Committee began issuing Requests for Applications (RFAs) on projects designed to improve the scientific understanding of the effects of motor vehicle emissions on human health. As illustrated in Figure 10.1,[4] the process of developing RFAs begins with a solicitation of views from both sponsors and the public. The Research Committee then identifies critical gaps in scientific knowledge and, through the issuance of the RFAs, invites the research community to submit proposals. Following review of the research designs and experimental protocols, the committee selects projects and awards contracts. The first award was made in 1983.

The Review Committee, HEI's other mainstay, began formal operation only in fiscal year 1985, the first year in which projects funded by the Institute produced concrete experimental results.[5] Maintaining a strict functional separation from the Research Committee, members of the Review Committee play no role in the design or conduct of the research they are called upon to review. When a study is completed, the Review Committee, assisted by HEI's in-house staff, makes a scientific evaluation of the research findings, identifies remaining uncertainties, and makes itself available as needed to interpret study results for the sponsors. The committee's role as a purely scientific arbiter precludes it from commenting on the possible application of the research to regulatory policy.

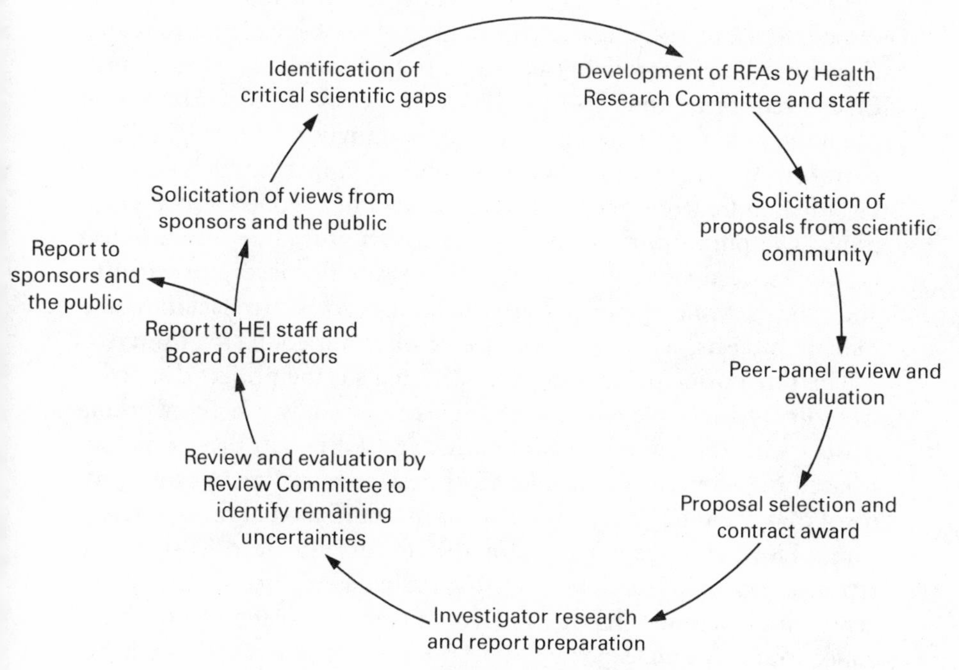

Figure 10.1. HEI research and report cycle.

HEI in Operation: The Multi-Center CO Study

HEI's most ambitious project in its first half-decade of existence was the Multi-Center Carbon Monoxide (CO) study, undertaken in 1983 at a projected cost of over $1 million.[6] The decision to fund this investigation was prompted by the troubled history of standard-setting on CO that has been described in Chapter 6. When the data on health effects supplied by Wilbert Aronow were thrown into question, EPA decided that similar studies conducted by reputable researchers were needed in order to determine the cardiovascular effects of CO exposure. The Aronow fiasco provided an ideal opportunity for testing the capabilities of the newly created HEI, and in July 1983 EPA transmitted to the Institute a formal request for a CO study. HEI's Research Committee, for its part, welcomed the chance to demonstrate that the organization could generate high-quality science even when the object was to serve an immediate policy goal. Designed to compensate for earlier failures, the project neatly illustrated the advantages and drawbacks of the Institute's unusual program for creating apolitical knowledge.

To launch the CO project, HEI created a small but distinguished Oversight Committee consisting of three members of the Research Committee, two cardiologists, and an environmental physiologist. EPA's need for reproducible results clearly required a departure from usual research-solicitation practices. Accordingly, the Oversight Committee itself designed a preliminary study plan that was submitted to several hundred prospective researchers with a request for applications. The purpose of this highly structured procedure was to select multiple investigators who could demonstrate the necessary scientific and nonscientific qualifications, including access to facilities and human subjects, as well as absence of bias and conflicts of interest.

The three principal investigators selected for the project decided it would be inadvisable to replicate the Aronow study exactly. Working closely with the Oversight Committee and HEI staff, they then confronted the monumental challenge of designing a more rigorous protocol that could be carried out in a consistent fashion at three different sites. How, for example, should the investigators ensure the repeatability of CO uptake rates for individual subjects? The multi-center study abandoned the traditional course of using identical CO concentrations and exposure times at each experimental exposure. Instead, the investigators adopted an alternative method in which CO concentrations were adjusted in accordance with each subject's uptake rate, a method deemed capable of generating statistically more powerful results.

But this approach encountered substantial obstacles. The process of creating a foolproof protocol turned out to be significantly more complex than HEI staff had expected at the outset. A number of technical problems proved "maddeningly resistant to resolution."[7] It became necessary, for example, to run a pilot study to ensure uniformity of uptake rates. According to Jane Warren, the study's project manager, "the effort has been to make the carboxyhemoglobin levels the same for each patient, and that's pretty complicated . . . it's a really messy area, trying to hit the CO level exactly. Much of the delay in the experiment has been caused by trying to reach these target levels."[8] Complexities such as these vastly increased the length and cost of the study. Work began in December 1983, but the first subjects were not tested until March 1985; by May 1987, when the study ended, some 69 subjects had been tested. HEI eventually spent $2.5 million rather than the initially expected $1 million on the CO study.[9]

HEI had been forewarned by the Aronow example that rigorous quality assurance during the study was indispensable to its ultimate acceptability. Responsibilities for quality assurance in the CO project were delegated to Arthur D. Little, Inc. (ADL), an organization with considerable experience in the general area of quality assurance and some familiarity with CO experimentation systems. Throughout the research, ADL staff scientists made periodic site visits to all three laboratories to ensure that experiments were being conducted in accordance with the protocol. Such monitoring does not come cheap; one former ADL expert estimated that quality assurance adds 10–20 percent to the total labor cost of a research project.[10]

Midterm assessments of the multicenter study generally reflected a spirit of cautious optimism rather than an expectation that the project's completion would have a decisive impact on policy. One of the principal investigators suggested that the study's primary accomplishment would be to raise the general level of research on air pollution and human health. Another commentator succinctly asked, "Will this be better knowledge to set policy? I think so . . . Will it be the bible? I don't think so."[11] EPA, too, indicated that it did not expect any single study, even one as meticulously designed and performed as the multicenter CO study, to be a scientific "silver bullet."[12] With regard to revisions of the CO standard, EPA believed it would be necessary to rely on a number of studies in progress, including those sponsored by HEI.

At HEI's fourth annual meeting in 1987, it became clear that the experiments were not proceeding precisely as planned. The investigators, for example, had experienced difficulty in finding suitable numbers of patients because of the strict criteria that had to be applied to ensure repeatability and because of therapeutic advances that had reduced the available population of candidate patients. There were also renewed questions about whether the measurement of carboxyhemoglobin levels in the study subjects was policy-relevant. Were these measures related to the effects on health, and could they be directly translated into an ambient air quality standard? If not, what was the utility of making these measurements? Although these questions might under other circumstances have blossomed into a full-scale deconstructive debate, the completed study eventually survived peer scrutiny both by the Review Committee and by the *New England Journal of Medicine,* where it was published in November 1989.[13] Study results confirmed the hypothesis that even low concentrations of CO can harm patients with heart disease, an outcome that tended to justify stricter standards and hence was potentially detrimental to industry's interests. That the study withstood challenge in spite of its possible impact on policy provides a favorable indication of HEI's ability to produce state-of-the-art science while operating within the constraints of regulatory politics.

Implications of the HEI Experience

Given HEI's short history, it is difficult to predict whether the CO success story will be repeated and, more generally, whether the HEI model will displace more traditional methods of producing and validating regulatory science. A 1986 investigation by the General Accounting Office (GAO) suggested that credibility problems at least were not going to be the Institute's primary concern. The sixty-one individuals interviewed in the course of the GAO study generally agreed that HEI had succeeded in maintaining its independence and was conducting high-quality research in accordance with its stated mission.[14] The interviewees, however, were less sanguine about the relevance of some of this work for policy. In a ranking of 100 HEI projects, EPA rated 44 as being of only "low" or "low-medium" relevance; the most frequent reasons for unfavorable assessments were that the studies were employing excessively high levels of exposure, and consequently were of limited utility for risk assessment, or that they were unnecessarily duplicative of earlier research.[15]

HEI's own staff interpreted these criticisms as the necessary price of success.[16] They noted that the Institute had been remarkably fortunate in attracting funds at a time of dwindling support for research on health effects. This track record set HEI in direct competition with EPA's in-house research units, whose own budgets were unavoidably cut back by grants to HEI. At the same time, HEI executives reacted defensively to the charges of irrelevance. It would be unrealistic, they suggested, for near-term policy payoffs of the sort expected from the multicenter CO study to serve as the sole criterion for evaluating the relevance of HEI's research projects. Institute staff members felt that their mission should be conceived in broader terms, asserting that HEI's business was to raise the standard of research pertaining to policy. HEI's ethos, in other words, was described in terms more typical of research science, where the pursuit of knowledge is valued for its own sake, than of the utilitarian world of regulatory science. Few regulatory agencies could afford to indulge in so long-sighted an orientation to scientific research.

The conflicts that surfaced in the wake of the GAO report were symptomatic of HEI's essential dilemma: is it possible to maintain scientific credibility, through expensive, high-quality research, while remaining relevant to policy? It is unlikely that HEI's sponsors would countenance many projects like the CO study unless they could be sure that they were gaining real advantages in terms of time, money, or reduced controversy. Meticulous protocols and perfect replicability are unlikely to seem cost-effective if the results do not bring regulators closer to making decisions. For example, if the CO study had proved contentious or deflected EPA from further rulemaking, the $2.5 million cost might well have been seen as excessive, even though it was but a small fraction of the $320 million that EPA had spent since 1980 on research to support air quality standards for criteria pollutants.[17]

A more immediate threat to HEI's credibility is the pressure, both internal and external, that the Institute is under to diversify its research agenda and its roster of clients. HEI's first major effort to meet these demands called attention to the delicate balance that the organization seeks to maintain between research science and regulatory science. In August 1988, Congress included in its appropriations for EPA a line item of $2 million for research on asbestos, which the legislators specified should be done by HEI in order to ensure the quality of the science and the objectivity of peer review. At the EPA administrator's request, HEI's board of directors convened a study team to recommend how the Institute should respond to the congressional mandate. The study team's first concern, understandably, was to assess the project's feasibility in scientific terms and to develop a research plan, but its inquiry also focused on the possible impact of asbestos research on HEI's institutional identity.

The study team recognized from the outset that the new project would have to be carried out in a politically charged atmosphere, marked by threats of litigation and continued uncertainty about the nation's policies for abating asbestos in public schools.[18] Within this context, the two institutional issues that most preoccupied the study team were, first, the high risk of civil liability associated with asbestos and, second, the possibility that research dealing with abatement strategies (perceived within HEI as regulatory science) might detract from the Institute's primary emphasis on basic research related to the health effects of air pollution. The solution the Institute adopted was, in a way, the ultimate example of boundary work: to create a new corporate entity, called Health Effects Institute—Asbestos Research (HEIAR), with its own director and scientific staff but governed by the same board as HEI. This organizational model, the study team felt, would both restrict the parent company's liability and protect it against too close an involvement in the regulatory politics of asbestos.

Although HEI's need to maintain a sharp boundary between research science and regulatory science played a key part in HEIAR's genesis, the daughter company, once established, was presented to the public as situated unambiguously on the research side of the boundary. Archibald Cox, chairman of the board for both companies, characterized HEIAR as "a fiercely independent and competent scientific body" that would replicate most of HEI's "proven approaches to seeking and reporting scientific facts to the public on complex and controversial issues of public concern."[19]

Risk Assessment without Politics

Straddling the gray area between science and policy, risk assessment remains the classic "trans-scientific" activity carried out by regulatory agencies. However, agencies themselves have occasionally delegated this function to independent scientific organizations, sometimes under specific instruction from Congress. Ironically, FDA, the agency with the best-developed reputation for in-house scientific competence, probably has the widest experience with such uses of outside expertise. In particular, FDA has turned to the National Academy of Sciences and FASEB for assistance with implementing newly tightened statutory standards for approving potentially hazardous drugs and food additives.

NAS and Drug Efficacy Review

After the passage of the 1962 Kefauver-Harris Amendments to the FDCA, FDA determined that some 4,000 "old" (that is, pre-1962) drugs should be reviewed for efficacy, although the statute did not expressly impose such a requirement.[20] To carry out the reevaluation, FDA contracted with the National Academy of Sciences. NAS established about thirty panels of six members each, including many experts who were not otherwise affiliated with either NAS or the National Research Council.[21] The panels were asked to rate the drugs on a six-point scale of effectiveness.[22]

In carrying out the reviews, the panels relied on the published scientific literature, information provided by FDA and manufacturers, and their personal judgments and experience.[23] While FDA provided general guidelines for the review, there were numerous inconsistencies in the criteria and procedures employed by various panels, and sometimes in their findings as well. Given the poor state of scientific knowledge about many of the drugs, moreover, a finding that a drug was "ineffective" often meant merely that there was insufficient evidence to judge its effectiveness.[24] Nonetheless, the monumental task was completed with surprising expeditiousness: in 1968 the NAS provided FDA with over 2,800 reports resulting from the Drug Efficacy Review.

While FDA's decision to consult with the Academy aroused no serious controversy, the agency's attempts to implement the recommendations of the review panels quickly ran into legal difficulties. A major point of contention was the summary revocation of a drug's approval and its removal from the market when FDA determined that there was no substantial evidence of efficacy. Drug manufacturers argued that taking such drastic action on the basis of the NAS-NRC panel findings was unfair. Industry spokesmen noted that the recommendations conveyed to the agency were unreviewed by the full Academy or Council and represented only the conclusions of the six-member panels rather than a larger scientific consensus.[25] Further, in view of the quality of the data and the time constraints placed on each panel, it seemed clear that many of the findings were more in the nature of educated opinions than of definitive scientific facts.[26] Some of the panelists evidently shared the drug industry's misgivings about the process, as the following brief apologia indicates:

> I was Chairman of the Psychiatric Drugs Panel in the Drug Efficacy Study. I wasn't unhappy with the panel, but I didn't like six people being in a position to adjudicate for the country, with all its varied opinions and patterns for practice . . . I didn't want the results used to regulate the practice of medicine.[27]

A number of manufacturers eventually brought lawsuits demanding more extensive hearings before FDA banned or restricted their products.[28] FDA countered with the argument that a hearing should be granted only when there was a question of fact about the existence of "substantial evidence." Industry's rejoinder that FDA had defined the substantial evidence standard too narrowly for old drugs failed to make much headway in the courts. Judges were clearly disinclined "to get 'too involved' in what they viewed as a morass of medicine, therapeutics and science."[29] The passivity of the courts simply confirmed the received wisdom among food and drug lawyers that it was useless to seek judicial review of FDA decisions, since courts automatically deferred to the agency's expertise.[30] The fact that FDA's judgments in this case were backed by the expert NAS-NRC panels made the case for deference all the stronger.

The Review of GRAS Substances

In its regulation of food additives, FDA relied on FASEB to perform a review similar to that conducted by NAS-NRC for pre-1962 drugs. As described in Chapter 8, FDA contracted with FASEB to ensure that substances "generally recognized as safe" (GRAS) prior to the enactment of the 1958 Food Additives Amendment were still safe in the light of current scientific knowledge. In so doing, the agency granted this professional society considerable discretion to structure the analytical framework for reviewing the scientific data relating to GRAS substances.

FASEB was formed in 1913 by three professional societies of biological and medical scientists[31] and has since grown to include three additional member societies and one affiliate member. Together, these organizations represent about 27,000 biomedical researchers. The Federation's purposes, as set forth in its constitution, include bringing together researchers from its member societies, disseminating research results, and otherwise furthering the group interests of the members. Since 1962 FASEB has become an important presence on the policy-making scene through its Life Sciences Research Office (LSRO). LSRO's primary function is to provide scientific advice to federal decisionmakers through a combination of critical literature reviews, consultation with knowledgeable experts, and assessments presented in the form of reports to agency officials.[32] While LSRO reports are often commissioned to aid the development of new research programs, agencies have also turned to LSRO for more focused studies connected with specific regulatory decisions. This was the role played by LSRO in managing the review process for GRAS substances.

To carry out the review of some 400 such food ingredients, LSRO convened a Select Committee (SCOGS) consisting first of nine and eventually of eleven members.[33] A number between ten and twelve was deemed most effective in order to allow for occasional inevitable absences at committee meetings. Committee members and their staff were selected so as to have no possible conflicts of interest, even though this criterion eliminated some of the "most knowledgeable and talented experts in the field of food technology and additives."[34] A suggestion that the committee should strive for balance by adding overtly "consumer-oriented" and "producer-oriented" members was rejected on the grounds that this approach might lead to overly adversarial proceedings.[35]

The complete GRAS review process (see Figure 10.2)[36] spanned the decade between 1972 and 1982. During this period SCOGS was forced to confront many previously undecided administrative, technical, and science policy issues relevant to the safety of foods and ingested substances. Questions concerning the appropriate forms for public participation and communication of the committee's findings were resolved as the committee gained experience and confidence. Thus, beginning with the twenty-seventh SCOGS report, FDA began announcing the availability of the committee's preliminary evaluations in the *Federal Register.*[37] The announcements stated that the committee would, at request, hold public hearings to receive data, information, or views pertaining to its draft report. The hearings were seen as a mechanism for ensuring that the views of divergent interest groups

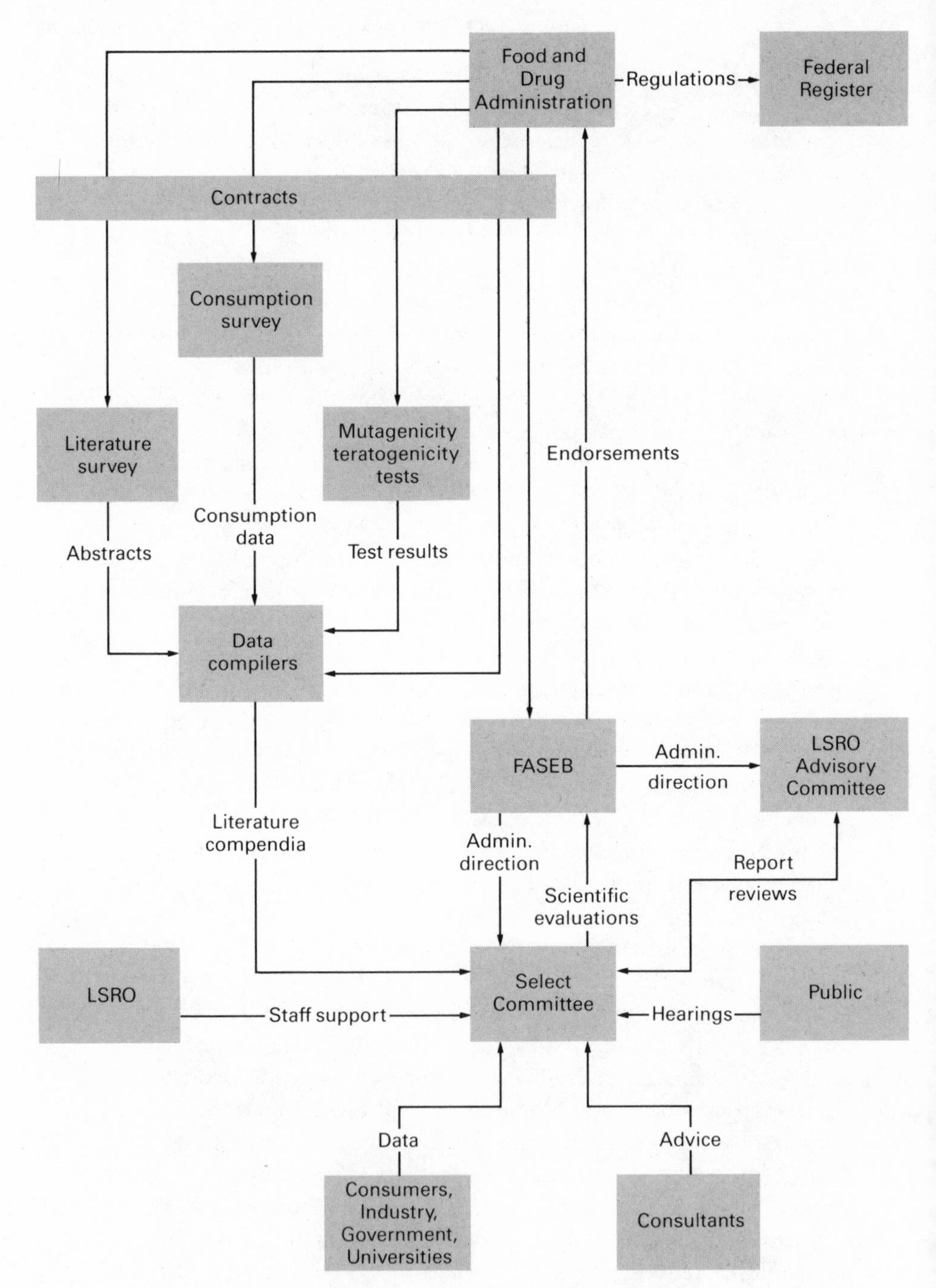

Figure 10.2. GRAS review process.

would be reflected in the final report without pitting these adversaries against one another. With respect to the risk assessments themselves, SCOGS soon realized that a number of alternative conclusions would be needed to characterize several hundred substances in an operationally useful manner. The committee, in consultation with FDA, formulated five different characterizations that were eventually applied to all but a handful of the evaluations. A major benefit of this approach, the committee felt, was that it reduced "the tendency of cautious scientists to qualify and 'write around' rather than make hard choices."[38]

The SCOGS review predated by almost ten years the NAS study that established the concepts of risk assessment and risk management in the policy literature. Yet members of the committee were well aware of the problematic nature of the boundary between their activities and the development of formal regulations by the FDA. The committee accepted as a working principle that regulatory policy lay beyond its purview, but the interim report issued in 1977 also acknowledged that "at times, considerations regarding the translation of scientific assessments into official regulations became inextricably imbedded within the very context of our legitimate concern."[39] The committee's 1982 report was even more forthright in calling attention to subjective elements that invariably enter into scientific evaluations related to policy:

> Among the principal sources of subjective variability among evaluators are: personal leanings concerning what constitutes "safety"; differences in perception of what constitutes adequacy of data by the same individual for different situations; the degree to which scientific popularity (the "conventional wisdom") is an influence; personal weighting of the significance of adverse findings based on unconfirmed studies and/or less than rigorous experimentation.[40]

Interestingly, however, the committee's primary suggestion for avoiding "an improper psychological and judgmental overlay" on scientific evaluations was to urge that experts should be "equipped to handle the extra-scientific factors impinging on their final options in a perspicacious and sound fashion." In other words, the committee saw broader social education of experts as the solution to problems of bias and subjectivity. This recommendation, of course, reflects an unabashedly technocratic view of policymaking; it is not surprising that a group of scientists should have come to such a position when considering how best to enhance their own role in the science policy process.

Because of its independence from FDA, the committee was able to comment on some of the most problematic aspects of food safety evaluation in ways that agency officials might not have been free to do. One such issue was the Delaney clause, about which SCOGS expressed serious reservations. In its 1982 report, for example, the committee suggested that judgments about carcinogenicity should be "particularized," with careful, case-by-case consideration of the dose range, diet, route, length, and circumstances of exposure.[41] This approach was inconsistent with the Delaney presumption that positive animal studies can be used as a reliable, if blunt, device for dividing all chemicals into carcinogens and noncarcinogens. In articulating a more cautious and scientifically shaded philosophy concerning the identification of carcinogens, SCOGS implicitly lent support to subsequent FDA efforts to break away from a literal implementation of that clause.

As in the case of the Drug Efficacy Review, however, SCOGS's greatest contribution to FDA's decisionmaking was to provide a relatively uncontroversial and efficient means for disposing of a heavy regulatory burden. Over a ten-year period, the committee was able to bring the basis for GRAS classifications scientifically up-to-date, thereby paving the way to eventual elimination of this safety category. Indeed, by 1982 the committee was in a position to recommend that GRAS substances and other food ingredients should be merged into a single system for purposes of assessing their possible health hazards.[42] In terms of the pace and contentiousness of the review process, both the Drug Efficacy Review and the SCOGS evaluation of GRAS substances thus stand in marked contrast to EPA's experience with the reregistration of pesticides under FIFRA.

The Public Board of Inquiry

FDA's experiment with the Public Board of Inquiry (PBOI) grew out of disenchantment with formal hearings as a mechanism for resolving scientific issues in rulemaking. Critics pointed to the length and inefficiency of formal proceedings and observed that adversarial procedures neither permitted the agency to consult with the nation's foremost experts nor were particularly well suited to the resolution of scientific issues.[43] The PBOI was conceived as a less legalistic alternative that would allow scientists to exercise greater control over scientific decisions. But although it was once regarded as a promising initiative, as of 1987 the PBOI had been used on only two occasions, once to evaluate the artificial sweetener aspartame and once in regulating the contraceptive Depo-Provera. A careful analysis of the PBOI for aspartame indicates why the procedure did not prove more appealing.

G. D. Searle, the manufacturer of aspartame, petitioned FDA for approval of the product in 1973, and FDA issued a proposed regulation permitting its use in 1974, but questions about its safety were raised by John Olney, a Washington University psychiatrist, and James S. Turner, a consumer advocate.[44] In addition, an FDA task force reported problems with Searle's research practices that cast doubt on test results for aspartame and several other products.[45] Administrative proceedings were stalled for several years while FDA assessed the accuracy and reliability of Searle's data on the sweetener. The agency contracted with UAREP, a nonprofit consortium of pathologists, to audit twelve of Searle's studies and to determine whether they had been carried out according to protocols and accurately reported to the agency. FDA reviewed another three studies in-house and concluded that all fifteen were reliable.

Both the manufacturer and the challengers thereupon waived their right to a full evidentiary hearing and opted instead to have the safety issues considered by a PBOI. A three-member board was appointed in 1979 from lists of nominees submitted by Searle, Olney, and the Bureau of Foods.[46] Olney objected to the board's composition on the ground that there was an overrepresentation of MIT scientists, leading to a possible conflict of interest, especially in view of the fact that the Bureau of Foods director and several Searle scientists shared the same academic background. In addition, Olney felt that the scientific qualifications of at least one board member in nutrition and metabolism were inappropriate and that the board should instead have included an expert in neuropathology. FDA, however, dismissed these objections as untimely and also because they failed to raise a substantial concern about the board's objectivity.[47]

The questions put to the board were negotiated in advance by the parties and included technical as well as policy elements. On the scientific side, the board was asked to determine whether the ingestion of aspartame "poses a risk of contributing to mental retardation, brain damage, or undesirable effects on neuroendocrine regulatory systems," and whether aspartame "may induce brain neoplasms in the rat."[48] Based on its findings with respect to these issues, the board was also asked to decide whether aspartame should be approved for use, and what labels, if any, should be required in connection with its permitted uses.

The PBOI held hearings in January and February of 1980 and rendered its decision in October. At this time, the board found no evidence that aspartame posed an increased risk of brain or endocrine dysfunction, but concluded that there was insufficient evidence to determine whether the substance could cause cancer. These concerns led the board to recommend that the product be banned pending further testing. In July 1981, however, the FDA commissioner overruled the board and approved the product for certain tabletop uses.[49]

Although none of the parties challenged this decision, FDA's subsequent broadening of the approved uses of the sweetener led to litigation. The Community Nutrition Institute (CNI) sued the agency for permitting aspartame to be used in carbonated beverages in a final rule issued in July 1984. CNI claimed that FDA had unlawfully denied it a formal hearing on evidence that was material to the agency's decision to approve the new uses of the product. The D.C. Circuit Court of Appeals, however, deferred completely to FDA's decision and to the agency's underlying judgments with respect to aspartame's health risks.[50] The court concluded that the new studies cited by CNI either had been properly considered and dismissed by the agency or else provided no new empirical evidence of harm. Accordingly the court held there was no obligation on the agency's part to grant the request for a hearing prior to permitting the use of the artificial sweetener in beverages.

Continuing its attack on aspartame, CNI petitioned FDA in 1986 to ban the product under the "imminent hazard" provision of the FDCA.[51] In support of this request the consumer group cited reports of more than eighty previously healthy young adults who had allegedly suffered seizures as a result of exposure to aspartame. FDA denied the petition in November 1986, and CNI appealed the ruling. In the meantime, CNI asked the agency to reconsider the earlier petition and objected to an apparent relaxation in FDA's standards for evaluating safety.

Following the marketing of aspartame FDA received a steady stream of consumer complaints describing adverse reactions, usually headaches and seizure symptoms, allegedly caused by the sweetener.[52] Continual monitoring and analysis of these reports, however, provided no grounds for the agency to suspect that aspartame was connected with any specific clinical syndrome or had harmed any identifiable groups of users. Sporadic scientific studies claiming to have found negative health effects also continued to appear in the literature. One staunch opponent of aspartame in the scientific community, Dr. Richard Wurtman of MIT, was accused by a colleague of an "obsessive" concern with the compound.[53] But these disputes within the scientific community neither caused FDA to rethink its position nor led to a court order compelling such rethinking.

FDA's success in withstanding scientific and consumer challenges to its aspartame decisions suggests that the PBOI in no way detracted from the agency's legitimacy. Yet the process remains a problematic mechanism for developing science policy. One feature of the board's decisionmaking that aroused particular misgivings was the fact that a panel of technical experts was asked to comment on both scientific and policy issues.[54] FDA's use of the PBOI appeared to confirm the fear that regulatory agencies in search of legitimacy would be tempted to abdicate policy authority to expert bodies having neither the competence nor the legal right to decide such matters. In the aspartame case itself the threat of undue control by experts did not materialize, since FDA overruled the board's policy judgment, just as the agency has done in other cases where it has deemed scientific advice to be too conservative.[55] Nonetheless, the open delegation of policy questions to an expert panel ran sufficiently counter to the science policy paradigm to convince some observers that the PBOI process should not be lightly repeated.

Others questioned the wisdom of a procedure in which FDA is totally free to accept or reject the findings of the board of inquiry. Clearly, this discretion makes it possible for the agency to manipulate the authority of science to its own advantage.[56] When the PBOI supports the agency's preferred policy position, FDA can rely on its opinion for additional support; at other times it can pay lip service to science by consulting the board but ignoring its recommendations. Seeking expert advice under such flexible ground rules does little to improve the quality of the agency's scientific determinations, though it may increase the already considerable potential for regulatory uncertainty and delay.

Finally, the procedural informality of the PBOI proved a source of unease for some, though not all, observers. While some viewed the board as creating an adequate record,[57] Nancy Buc, a former chief counsel for FDA, rendered a much harsher judgment:

> the PBOI has created an utter mess. The questions about whether their own views are evidence, their viewing of the slides, not putting

> anybody under oath, not having a record, not even citing one single source in the record in their opinion is a source of rather considerable aggravation to the lawyers who are then going to have to make something out of it that certainly will be reviewed by a court.[58]

Wider Applications

How should HEI's and FDA's experiments in technocratic decision-making be interpreted in the context of the overall debate about regulatory science? More specifically, do the cases considered in this chapter provide a sufficient rationale for fundamentally reshaping current processes of research funding and rulemaking? In two of the three cases—EPA's involvement with HEI and FDA's with NAS and FASEB—there appears to be ample evidence that use of novel institutional and procedural arrangements gave the agencies real advantages in terms of increased efficiency and perhaps enhanced scientific legitimacy. The PBOI case is more ambiguous, although this procedure, too, arguably provided a net benefit for FDA. Yet each mechanism contains flaws that suggest a need for cautious generalization. Each is best suited to addressing only a specialized subset of the problems engendered by regulatory science, and none can be regarded as a panacea for regulators confronted by scientific controversy.

The HEI model provides, first and foremost, a promising alternative to the generation of policy-relevant science by politically motivated organizations. The Institute's public-private financing arrangements and its independent committee structure seem ideally calculated to reassure the users of regulatory science that the biases reflected in HEI-funded research are not those of government, industry, or any other party with an interest in the political application of knowledge. Relatively insulated from the deconstructive pressures of regulatory politics, HEI can facilitate the informal micronegotiations that are so necessary to building a robust scientific consensus. The experience of designing the CO multicenter study suggests, however, that even for generating new data the HEI model should be used sparingly and only when certain prior conditions are met. The need for a particular study or research direction should be acknowledged by all relevant actors and existing data should be clearly recognized as qualitatively or quantitatively inadequate. Accordingly, a prehistory of unresolved scientific conflict may be a desirable, though not necessary, prerequisite for HEI's intervention. Finally, the expense of the research should in some way be commensurate with the importance of the regulatory issues confronting HEI's sponsors.

The controversies discussed in earlier chapters suggest that these conditions are present frequently enough to justify the existence of research-sponsoring institutions that are independent of regulatory agencies. The disputes over epidemiological studies at Love Canal and among formaldehyde workers, for instance, might have been averted if the research had been funded and peer-reviewed through a process like HEI's. At the same time, it is unclear to what extent the present HEI structure permits a reliable balancing of such disparate factors as the inherent interest of the research, the saliency of the related policy issue, and the project's capacity to satisfy regulatory adversaries. It remains, in any case, an open question whether the Institute can permit pragmatic and political considerations to influence its research agenda without being tainted by the appearance of interest. The formation of HEIAR graphically illustrates how the strategic use of boundary work can help address the problem.

The role played by NAS and SCOGS in reviewing drugs and food additives is particularly interesting because it conformed more to a European than an American pattern of decisionmaking. It is no coincidence that both bodies were enlisted by FDA, the U.S. agency whose history and mission have most closely approximated those of expert agencies in other industrial countries. Like many European advisory committees, the NAS and SCOGS panels operated with relatively informal procedures that were consultative rather than adversarial. Their mandate ostensibly was to assess only technical issues related to risk and safety, but they delivered judgments that included significant components of policy as well as science. Finally, the experts selected by NAS and SCOGS were quite effective in persuading FDA to adopt their recommendations as policy.

But does FDA's success with delegations to professional societies imply that this model should be widely adopted in decisionmaking based on regulatory science? Did the National Research Council err in recommending that risk assessments should not as a general matter be delegated to independent scientific committees? To draw positive responses based on FDA's experiences alone would be premature indeed. Both NAS and SCOGS were confronted by relatively atypical regulatory situations. Although hundreds of products had to be brought into conformity with new statutory standards, there was no evidence that any were posing an immediate threat to public health or safety, and FDA was not under pressure to act within a specified time frame. The expert-committee decisions were cumulatively weighty, but individually of limited economic and political significance; technically, moreover, they were merely recommendations to FDA. Under the circumstances, the advisory panels could carry out their tasks deliberatively, incrementally, and almost invisibly. Such a posture would be virtually impossible to maintain if independent expert committees were transformed into the primary locus for decisions involving a mix of science and policy, especially in the current era of distrust for government and heightened public awareness of risk.

The PBOI marks the least radical departure from existing regulatory practice as well as the one whose utility is most questionable. In replacing trial-type administrative hearings, the PBOI may well shorten the time to reach final decisions, although the scanty history of this procedure does not allow a definitive evaluation of this point. The inquiry on aspartame, however, raises troublesome questions that must be considered if attempts are to be made to adopt the model on a wider scale. In particular, the extent of the agency's discretion to overrule the board on its technical findings should be clarified. If agencies retain complete discretion in this regard, then consultation with the board could easily become a counterproductive exercise, creating yet another focal point for contention and adding to the complexities of an already overloaded administrative process.

The most significant conclusion to be drawn from all three examples considered above is that the democratic and technocratic forms of decisionmaking maintain a creative dialectic in the U.S. regulatory process. Both EPA and FDA, it appears, are willing to tolerate departures from the "democratic" science policy paradigm in favor of more expert-centered processes, though usually in connection with narrowly defined regulatory objectives. In seeking to avoid scientific controversy, agencies can clearly resort to a wide variety of institutional and procedural alternatives, spanning the continuum from full democratic participation to specialized expert deliberation. They range from informal rulemaking, with or without scientific advice, to formal contracts with the National Academy. The challenge for regulatory reform is to determine where in this continuum science policymaking should be situated under particular scientific, legal, administrative, and political circumstances. To address this question in the light of cases considered throughout this book is a task for the concluding chapter.

11

The Political Function of Good Science

Regulatory practices at EPA and FDA would seem to indicate that the technocratic vision of public policy has scored important gains over the competing democratic paradigm. Scientific advice has become an integral part of decisionmaking at both agencies. Both are served by, and are accountable to, an impressively diverse array of ad hoc and standing expert committees. There is growing awareness in both agencies that timely consultation with outside experts can prevent controversy or, at the very least, protect effectively against challenge. Both recognize, too, that their own reputation for scientific excellence hinges on their ability to maintain productive collegial relations with the external scientific community.

Yet the picture that emerges from a close scrutiny of the advisory process does not look wholly reassuring from either a technocratic or a democratic standpoint. The contingent and socially constructed character of regulatory science challenges conventional technocratic assumptions about the nature of scientific knowledge and the role of experts. Advisory committees, we know from experience, rarely restrict their deliberations to purely technical issues. In fact, the experts themselves seem at times painfully aware that what they are doing is not "science" in any ordinary sense, but a hybrid activity that combines elements of scientific evidence and reasoning with large doses of social and political judgment. But if science is missing or obscured in the advisory process, scientists in the aggregate wield influence, and they do so, moreover, through proceedings that lack many of the safeguards of classic administrative decisionmaking. Participation by lay interests is limited and often one-sided, cross-examination is almost unknown, and committee recommendations, however much weight they carry, are seldom accompanied by detailed explanations or consideration of alternatives.

In the light of these findings, our earlier questions about the legitimacy of science-based decisions demand careful reconsideration. Must we accept David Collingridge and Colin Reeve's pessimistic conclusions about the power of science to rationalize policy, and does that view, in turn, lead to the cynical corollary that scientific advice is merely a thin disguise for the transfer of policy authority to experts? How can the notion of "improving" the scientific basis for regulation, a holdover from a discredited positivistic model of scientific knowledge, be reconciled with an analysis that emphasizes the negotiated and contingent dimensions of regulatory science? And how important, finally, are traditional forms of public participation in the advisory process? Is it desirable for expert committees to involve multiple political viewpoints in their deliberations? More generally, do procedures for advice-giving bear in any meaningful way on the ultimate social acceptability of policies based on esoteric and incomplete knowledge?

We begin this chapter by reexamining the connections between scientific advice and policy formulation and the strategies by which expert advisers achieve technical and political legitimacy. A point that emerges forcefully from the documented cases is that scientific advisory proceedings—no less than administrative proceedings of a nontechnical kind—are most effective in building consensus and guiding policy when they foster negotiation and compromise. The definition of "good science" by advisory committees emerges in this light as an important adjunct to the more openly political negotiations that underlie the development of regulatory policy. In the concluding sections of the chapter, we explore the implications of this analysis for future relations among agencies, their scientific advisers, and the public.

From Advice to Policy

Although pleas for maintaining a strict separation between science and politics continue to run like a leitmotif through the policy literature, the artificiality of this position can no longer be doubted.[1] Studies of scientific advising leave in tatters the notion that it is possible, in practice, to restrict the advisory process to technical issues or that the subjective values of scientists are irrelevant to decisionmaking. The negotiated and constructed model of scientific knowledge, which closely captures the realities of regulatory science, rules out the possibility of drawing sharp boundaries between facts and values or claims and context. While policymakers may not openly subscribe to this view of science, it is worth noting that they, too, have retreated from a rigid support for separatism. A new flexibility is apparent in EPA's approach to scientific advice, with the result that advisory committees now have significantly more say in that agency's science policy determinations.

Evidence from regulatory case histories suggests, further, that proceedings founded on the separatist principle frequently generate more conflict than those which seek, however imperfectly, to integrate scientific and political decisionmaking. The example of the ozone rulemaking remains especially striking in this regard. In the first NAAQS revision, EPA staff members jealously guarded their prerogative to make science policy determinations such as the definition of "adverse health effect." This attitude, however, led to an unmanageably adversarial relationship between the agency and its expert advisers. The SAB committee became convinced that EPA was encroaching on matters that should have been left to scientists, and this view was eagerly exploited by the agency's industrial critics. By the time the ozone NAAQS came up for a second revision, EPA had adopted a much more conciliatory posture. CASAC members now were free to comment on any issues they deemed important at the borderline of science and policy, and their final dispositions with respect to these matters were markedly more supportive of EPA.

By contrast, confrontationally structured peer review in the Alar case was much less effective in resolving the scientific debate. The SAP initially imposed its own evidentiary standards on EPA, but at the cost of becoming too closely identified with industry's interests. The perception that EPA and its technical advisers had been captured by industry fueled NRDC's successful efforts to project an independent account of risk to the public. Alar eventually was withdrawn from the market, but public confidence in EPA suffered, and many scientists were left with the impression that the media and the environmental activist community had illegitimately seized control over the definition of good science.

FDA's proceedings on sulfites and aspartame offer similar contrasts. The ad hoc panel on sulfiting agents was formally charged with reviewing new medical information on sulfites and reevaluating the GRAS status of these compounds. The scientific issues, however, imperceptibly shaded into policy, as the panel recommended that warning labels be used to protect specially sensitive individuals. Interestingly, this breaching of the boundary between science and policy proved to be relatively uncontroversial. The inclusive nature of the hearing, the blurring of scientific and political authority, and the panel's own flexibility defused tensions that might have persisted in a more confrontational setting. The aspartame PBOI, too, intertwined risk assessment with risk management, but here the choice of process threw into sharp relief the panel's crossover from science into policy, producing an unstable decisionmaking environment. Critics rightly complained that a science court was an inappropriate forum for deciding questions of regulatory policy. Ironically, as well, the PBOI's quasi-judicial status attracted notice and sparked questions about its disciplinary breadth and technical competence, thus detracting from its scientific authority. These charges so weakened the Board that FDA was ultimately at liberty to disregard its findings as a basis for policy.

Acceptable Risk

Decisions whether or not to accept a certain level of risk are generally regarded as involving both personal and social values,[2] and hence as inappropriate for delegation to experts. In practice, however, advisory committees are often drawn into determining acceptable risk, particularly when making threshold decisions about the adequacy of evidence. This determination almost invariably entails a balancing of science against safety. Is the risk severe enough to warrant immediate regulatory action, or should one wait for more and better data? Though the answer may be couched in technical terms, as a verdict about the adequacy of the evidence, it necessarily incorporates a sociopolitically colored judgment about the acceptability of risk.

The Alar case clearly conformed to this pattern. SAP's function was ostensibly to review the technical validity of OPP's determination that daminozide and UDMH, daminozide's breakdown product, presented a significant risk of human cancer. But in recommending that OPP should wait for additional studies, the panel delivered a collective judgment that the risks of another four years of exposure to these compounds were not serious enough to justify an immediate ban. In effect, then, the committee made a risk-benefit determination of the kind that FIFRA on its face delegates to the EPA administrator. EPA's acquiescence in SAP's judgment testified to the potency of such an expert consensus. Though the advisory committee strayed beyond its expertise, the weight of its scientific authority kept policymakers from disregarding its evaluation of Alar's risk. NRDC, by contrast, launched a successful collateral attack on the committee's findings by stressing not only the magnitude of the hazard but its possibly disproportionate impact on children. This strategy worked brilliantly precisely because it *re*politicized a supposedly technical issue by highlighting—or, to use a theatrical metaphor, foregrounding—its subjective, emotive, and political constituents.

FDA's advisory committees likewise have great power to decide whether a given body of evidence establishes safety or justifies regulatory intervention. In making such determinations, committees necessarily weigh the risks and benefits of alternative regulatory strategies, although this analysis may never be made explicit except in closed committee meetings. Thus, when the sulfites panel decided that the reports of adverse reactions were sufficiently serious to merit regulatory concern, it implicitly balanced the risks to asthmatics and salad-bar users against the benefits to society of continued sulfite use and more definite information about causal connections. The risks in this case were adjudged more significant than the benefits. The aspartame PBOI similarly determined that approval of the sweetener was too risky given the uncertainty of the evidence. Some of FDA's drug advisory committees acted in a comparable spirit in demanding stronger evidence of safety before clearing a drug for a new use or indication.

In a somewhat less transparent way, the OSTP risk-assessment document also illustrated the power of scientific advice to influence the definition of acceptable risk levels. Many of the principles of cancer risk assessment endorsed by OSTP were previously well entrenched as principles of administrative decisionmaking. For example, every federal risk-assessment policy beginning with EPA's cancer principles accepted the proposition that animal carcinogens should, in the absence of persuasive counterevidence, be regarded as human carcinogens. Agencies, in other words, saw the risk of cancer as severe enough to justify reliance on indirect evidence from other mammalian species. However, it was OSTP's effort to gather a substantial scientific consensus around these principles that finally gave them sufficient weight to withstand political challenge. In EPA's confrontation with OIRA over the risk-assessment guidelines, the regulatory agency prevailed largely because it could fall back upon the scientific consensus built by OSTP and further ratified by the SAB.

Thus, a key set of administrative policies for determining acceptable risk assumed political authority only after extensive review by outside experts certified them as adequately "scientific."

Scientific Advice as Legitimation: Negotiation and Boundary Work

What emerges from a successful recourse to scientific advice, then, is a very special kind of construct: one that many, perhaps most, observers accept as science, although it both shapes and is shaped by policy. That such constructs sometimes break down under political pressure is hardly surprising. Their frequent durability is the greater puzzle, for they are founded neither on testable, objective truths about nature, as presupposed by the technocratic model of legitimation, nor on the kind of broadly participatory politics envisaged by liberal democratic theory. To explain this puzzle, we turn to two microprocesses familiar both from regulatory case histories and from the social studies of science: negotiation and boundary work.

Regulatory practices at EPA and FDA support the thesis that negotiation—among scientists as well as between scientists and the lay public—is one of the keys to the success of the advisory process. The importance of incremental adjustments among divergent scientific viewpoints was clearly seen in the evolution of the federal guidelines for cancer risk assessment and in the equally troubled regulatory history of formaldehyde. Both cases began with the interested parties widely separated, conceptually as well as politically, but the differences lessened over time as the parties were gradually drawn into the negotiating process. For issues on a smaller scale, even a single proceeding may suffice to create scientific closure, provided it brings together in a negotiating format the majority of parties with a stake in the policy outcome. FDA's sulfite proceedings and some CASAC reviews testify to this point.

Contrary to the presumptions of the technocratic model, negotiation among scientists alone is not sufficient to ensure the public and judicial acceptability of agency decisions based on regulatory science, even if the experts in question represent a wide range of disciplinary and institutional affiliations. The protracted controversy over formaldehyde makes a powerful case that, when the stakes are high enough, no committee of experts, however credentialed, can muster enough authority to end the dispute on scientific grounds. The broadly representative federal panel on formaldehyde, for example, provided CPSC no tangible support in fending off an industry-initiated lawsuit, just as the use of a similar panel by FDA failed to persuade the D.C. Circuit that the agency was correct in its reinterpretation of the Delaney clause. Other efforts, such as the ambitious, scientifically pluralistic, formaldehyde consensus workshop, were equally ineffectual in establishing definitive interpretations of the evidence for use in making policy.

It could be argued that SAB's review of EPA's formaldehyde risk assessment built a robust scientific consensus without intervention from political actors. This procedure, however, was woven into a lengthy rulemaking process that included many prior attempts to define the scientific status quo. Accordingly, although EPA's credibility was not seriously challenged after publication of the revised risk assessment, it is unclear how much of the credit can be assigned to SAB's peer review. By the time the SAB looked at EPA's risk assessment, the scientific dossier on formaldehyde was much larger than in 1980, when CPSC first attempted to ban UFFI. Accumulating epidemiological evidence had persuaded most observers that formaldehyde was not an especially potent human carcinogen. At the same time, referral of the textile-worker issue to OSHA had lowered the formaldehyde industry's stakes in the outcome of EPA's rulemaking. With OSHA clearly responsible for regulating occupational exposures, any risk assessment carried out by EPA carried fewer consequences than at the time of the initial TSCA Section 4(f) controversy. Finally, industry's technical ability to control formaldehyde had improved to the point where many companies either voluntarily reduced exposure or left the market altogether. In the face of these developments, it would be unrealistic to insist that scientific peer review was the most important factor in defusing the formaldehyde debate.

While the formaldehyde case illustrates the importance of "ripeness"[3] in negotiations over regulatory science, EPA's relations with CASAC and SAP underscore the importance of procedural regularity. In the first ozone rulemaking, for example, EPA's evasion of its designated advisory body proved politically costly. The short-term advantage conferred by the Shy Panel was more than offset by the negative publicity generated by the use of this ad hoc and seemingly biased advisory body. Similarly, the dissolution and reconstitution of the SAP under an anti-environment administration broke the continuity of its relations with EPA and contributed to the impression that the panel was captive to business interests. In turn, OPP's own departure from established administrative practice in seeking expedited action on Alar damaged its credibility with the advisory panel.

If negotiation is the engine that drives the construction of regulatory science, boundary work is the casing that gives the result legitimacy. Boundary work by scientists grows out of a premise that seems diametrically opposed to the concept of negotiation and yet is equally essential to the closure of controversy. By drawing seemingly sharp boundaries between science and policy, scientists in effect post "keep out" signs to prevent nonscientists from challenging or reinterpreting claims labeled as "science." The creation of such boundaries seems crucial to the political acceptability of advice. When the boundary holds, both regulators and the public accept the experts' designation as controlling, and the recommendations of advisory committees, whatever their actual content, are invested with unshakable authority.

Scientific advisory proceedings, as we have seen, provide many opportunities for boundary work by agencies as well as their advisers,[4] and committees on the whole have been quite successful in designating as science any issues on which they wish to exercise influence. Curiously, however, the most politically successful examples of boundary work are those that leave some room for agencies and their advisers to negotiate the location and meaning of the boundaries. Examples include FDA's negotiations with its drug advisory committees about the application of legal as opposed to scientific standards of evidence and EPA's gradual softening of its conceptual boundaries in its relations with the SAB and CASAC. The rigid boundaries created by the SAP in the dicofol and Alar cases, by contrast, were easily overwhelmed by NRDC and the discussion of science was moved to a populist forum—television—much to the distress of those who wished to restrict the debate to more expert-dominated settings.[5]

Defining "Good Science"

The picture presented above of the advisory process is markedly at odds with the simple technocratic paradigm of "speaking truth to power," although, paradoxically, not inconsistent with it. According to this account, the committees attached to EPA and FDA do indeed help the agencies define good science—and this consensus view of science in turn influences policy—but they perform this function in part through skilled boundary work and in part through flexible role-playing. Protected by the umbrella of expertise, advisory committee members in fact are free to serve in widely divergent professional capacities: as technical consultants, as educators, as peer reviewers, as policy advocates, as mediators, and even as judges. Though their purpose is to address only technical issues, committee meetings therefore serve as forums where scientific as well as political conflicts can be simultaneously negotiated. When the process works, few incentives remain for political adversaries to deconstruct the results or to attack them as bad science. This stabilizing impact of expert advice can be observed at four critical junctures in the evaluation of regulatory science: validation of long-term research strategies; certification of study protocols and analytical methodologies; definition of standards of adequacy for scientific evidence; and approval of inferences from studies and experiments. Admitting the scientific community into each of these areas of decisionmaking produces a stronger consensus than any achievable through the agency's in-house expertise alone.

Research Strategies

One of the least visible activities of advisory committees is to guide the development of agency research programs. Scientific advice is here furthest removed from regulatory policy. Yet it is arguably in this area that the agency-advisory relationship produces some of its most beneficial results. In EPA's case, for example, both SAB and CASAC played a constructive role in reorienting the agency's research priorities toward greater attention to basic research on long-range environmental problems. Over time, such advice has positive, though diffuse and delayed, consequences for the credibility of regulatory science. It decouples the production of knowledge from immediate political pressures, thereby bringing the goals of regulatory science (see Chapter 4) into closer alignment with those of research science. This rapprochement, in turn, helps to solidify the agency's reputation for expertise and, with it, the overall scientific credibility of regulation.

Protocols and Analytical Methods

Just as peer reviewers generally cannot detect fraud or fabrication in scientific publication, so advisory committees will fail to uncover most instances of intentional misrepresentation in regulatory science, such as the CO studies submitted by Wilbert Aronow to EPA.[6] The expert committees attached to EPA and FDA do not normally review laboratory notebooks or other raw data, where problems in data generation can most easily be detected, and they have neither authority nor opportunity to carry out routine inspections of research facilities. Removed from the production of data, advisory committees are in no position to substitute for effective inspection and data-audit programs. At best, a committee may be able to render a secondary judgment about the adequacy of an agency's quality control efforts, as in the debate over the Toth studies in the daminozide case.

The scientific reputation of regulatory agencies, however, suffers far more frequently from charges of error, or "sloppy science," than from allegations of intended fraud. Conflicts over research design or analytical methods are especially likely to come to the fore in the advisory process, since problems of this type generally leave traces in the documentation provided to advisers. Thus, external advisory panels called attention to the absence of contemporaneous controls in the Love Canal study and to the failure to record times and levels of exposure in the Alsea basin study of 2,4,5-T.

Such after-the-fact attributions of error may protect agencies against unwise regulatory initiatives, but they have also proved to be both wasteful and divisive, and advisory committee resources are unquestionably better used to prevent such conflicts through careful prior negotiation of research design and methods. The tightening of peer-review procedures and good laboratory practice rules, as well as the delegation of research to the Health Effects Institute, reflects a growing commitment by both EPA and FDA to involving independent experts in the design of potentially controversial studies, thereby forestalling debacles of the kind described in Chapter 2. Payoffs from these efforts include the work of CASAC consultants in helping EPA refine the methodology of judgmental probability encoding and of the Cardio-Renal Committee in advising FDA with respect to clinical trials for antiarrhythmic drugs. Despite these improvements, however, regulatory peer review remains only a partial prophylactic against methodological disputes. Unless the review process successfully negotiates underlying political divergences, it is unlikely to protect a study from subsequent deconstructive challenges or, in Collins's terms, experimenters' regress. The fate of the intensively reviewed NCI study of formaldehyde workers speaks to the difficulty of building consensus around an issue of regulatory science after it has become politically contentious.

Evidentiary Standards

In its exploration of the relationship between regulatory policy and scientific fact, the *Ethyl* court suggested that claims not yet accepted as valid by the scientific community could nevertheless serve as a basis for regulation.[7] The opinion implied that agencies with a mandate to protect public health and the environment could draw upon their reserve of policy authority to certify scientific claims as suitable for use in policymaking. The court displayed a sophisticated understanding of technical uncertainty and the constructed character of knowledge, but it failed to appreciate that scientists as well as political decision-makers would have to be involved in negotiating claims at the boundary of science and policy. The opinion went astray in suggesting that "trans-scientific" questions, such as the adequacy of evidence, could be committed exclusively to the discretion of regulatory agencies.

The institutionalization of scientific advice, especially at EPA, has helped to rectify this oversight. Expert advisory committees now play a pivotal role in determining whether the evidence before the agencies is adequate for purposes of regulation. In the process, they may apply different and stricter standards to the data than are deemed necessary by the agency's in-house staff. During the Alar review, for instance, OPP determined that the Toth studies could be relied on for quantitative risk assessment, while the SAP, subscribing to the methodological criteria proposed by Uniroyal, held that evidence from this source was virtually useless ("not science").

The SAP in this case chose not to negotiate with EPA, but to impose on the agency a possibly excessive professional bias in favor of high levels of certainty. The fear that the scientific community will consistently pursue this strategy, however, does not find uniform justification in regulatory case histories. In reviewing the criteria document for carbon monoxide, for example, CASAC concluded that there was sufficient support for the existing air quality standard, even though some studies relied on by EPA were inconclusive and others were fraudulent. The committee's acceptance of the existing record, despite its inadequacies, reflected a willingness to temper scientific considerations with concerns for public health. The Federal Panel on Formaldehyde displayed a similar public health bias when confronted by insufficient data; the panel's conclusion that formaldehyde should be presumed to pose a carcinogenic risk to humans was consistent with a preference for Type II errors (that is, regulating substances later shown to be harmless).

The expected scientific bias toward rigorous proof is perhaps most frequently encountered in the work of FDA's advisory committees, but unlike EPA in the Alar proceeding, FDA has not always felt constrained to abide by its advisers' evaluations. The standards by which the aspartame PBOI evaluated the toxicological data, for instance, were later dismissed as inappropriate by FDA. Similarly, in the propranolol and Triazure reviews, the agency disagreed with its advisers as to whether there was substantial evidence of efficacy to satisfy statutory requirements. In testimony to the Fountain committee, FDA officials, as we saw, resorted to boundary work to explain these divergences. Their claim was that the task of fitting the evidence to statutory standards belonged finally with the agency and should be carried out in accordance with legal/administrative rather than scientific/technical standards. The agency's successful defense of this position can be attributed largely to the fact that FDA's evidentiary standards led in these cases to more permissive policy results—results less likely to be challenged by industry—than were recommended by the agency's scientific advisers.

Validating Inferences

The interpretation of evidence is an increasingly important focal point for scientific advice, as regulatory agencies turn to their expert committees to certify that their conclusions are reasonably derived from the data. Questions concerning the agency's reading of scientific data are heard more frequently in EPA's advisory proceedings than in FDA's, corresponding to the fact that FDA rarely confronts advisory committees with its own independent interpretations of regulatory science. Even EPA's advisory committees, however, very seldom charge the agency with outright error in its scientific inferences, a tribute no doubt to the developing sense of cooperation between experts and agency staff. More commonly, the advisory function is geared to making sure that the agency has examined the relevant data, asked the appropriate critical questions, and couched its technical arguments in a clear and consistent manner. CASAC's supervision of the ozone criteria document and SAB's review of the guidelines for cancer risk assessment and the formaldehyde risk assessment perfectly illustrate this dynamic. In each of these cases, the advisory committee alerted the agency to problems of presentation and argument, suggesting ways in which confusions and inconsistencies might be cleared up.

As long as such review remains nonconfrontational, it certifies that the agency's scientific approach is balanced and rational and that its conclusions are sufficiently supported by the evidence. Conscientious scientific review, in other words, serves many of the same functions as judicial review. Indeed, the questions posed to advisory committees by agencies closely parallel those that litigants have traditionally posed to the courts. Is the analysis balanced? Does it take account of the relevant data? Do the conclusions follow rationally from the evidence? Is the analysis presented clearly, coherently, and in a manner that is understandable to nonspecialists? To answer these questions competently, advisory committees are required to take a "hard look" at agency decisions. If they remain unsatisfied, they may even "remand" the case to the agency for additional procedures leading to a better record (for example, SAB's recommendation that EPA consult a pharmacokinetics review panel for formaldehyde). Approval by an advisory committee, finally, is equivalent to a judicial determination that the agency action is in compliance with applicable standards of substantive rationality. We will return below to the procedural implications of this convergence between judicial and scientific review.

Normative Implications

The ultimate goal of policy analysis is to bridge from the empirical and analytic to the prescriptive. Yet this essential linkage is all too seldom attempted in works that seek to understand the nature of scientific claims and their sources of authority. Much of this book has been devoted to analyzing the processes by which advisory committees certify the validity of claims arising from regulatory science. We have established the centrality of such phenomena as negotiation, construction and deconstruction, and boundary work. It remains now to apply this conceptual framework to concrete proposals for restructuring the relations of regulatory agencies, the scientific community, and other social and political actors. What general prescriptions should agencies follow in securing expert involvement in regulatory decisionmaking? Conversely, are there procedural or institutional arrangements that should be avoided in establishing an external advisory system?

Mandatory versus Discretionary Advice

Whether scientific advice should be mandatory or discretionary has long been seen as a central question in debates about science and regulatory policy. The thrust of the preceding analysis suggests, however, that this may be a misdirected concern. The primary rationale for the advisory process is, as I have argued, to engage the scientific community in negotiating a consensus over regulatory science. The science policy paradigm of the 1970s failed largely because it did not adequately emphasize the need for such involvement by independent experts. But as long as science is institutionally well represented in the regulatory process, it should matter relatively little whether advice is sought pursuant to legal mandates contained in statutes such as FIFRA and the Clean Air Act or in accordance with established agency policy, as at FDA. Provided their interactions are regular and predictable, agencies and the expert community can be expected to develop enough confidence in each other's motives and competence to negotiate productively on scientific as well as policy differences.

There are obvious advantages to be gained from preserving some maneuvering room in the way agencies structure their relations with outside experts. As EPA's interactions with the SAB and CASAC demonstrate, it is useful for agencies to tailor their consultative processes incrementally to fit the demands of particular regulatory programs. CASAC's two-step involvement in reviewing the NAAQS criteria document and staff paper represented the endpoint of a learning process to which both the agency and the committee contributed and which neither could have foreseen at the beginning of their relationship.

Discretionary advisory systems, however, have one drawback that should not be overlooked. If agencies are free to determine when to seek advice, the decision to consult or not consult can be used as a powerful tactical weapon to delay decisions and justify inaction. Delegating a sensitive issue to an advisory committee remains one of the most politically acceptable options for regulatory agencies, even when the underlying motive is to transfer a fundamentally political problem to the seemingly objective arena of science. FDA has frequently been criticized for using its committees in this manner, and the cases of Zomax, sulfites, and color additives lend support to the charge. In the long run, however, attempts to bypass established review processes (nitrites at FDA, ozone at EPA) appear to be more detrimental to an agency's credibility than delays occasioned by unnecessary consultation.

Committee Membership: The Issue of Balance

Given the importance of negotiation and consensus-building in the advisory process, it is clear that the ideal adviser needs to be more than a mere technical expert. This indeed squares with the perceptions of regulatory insiders, both scientists and administrators, who warn against undue narrowness in advisers. The most valued expert is one who not only transcends disciplinary boundaries and synthesizes knowledge from several fields but also understands the limits of regulatory science and the policy issues confronting the agency. Without this kind of breadth, scientists are likely to feel out of place in multidisciplinary committees and may not recognize the importance of negotiating standards of quality and certainty in regulatory science.

EPA's practices in appointing scientists to the SAB illustrate the affirmative actions that agencies can take to ensure that advisory committee members display appropriate breadth and balance. SAB members are often tested through a variety of focused appointments, such as consultancies on specific issues and membership on ad hoc subcommittees, to see whether they can function usefully in the larger theater of the full advisory board. This kind of informal testing and training is imperative to the efficacy of an advisory program and could be impeded by overly stringent requirements that advisers be drawn from particular institutions or from lists recommended by organizations such as NAS.

Agency advisers, as we have seen, must be capable of doing convincing boundary work while they are engaged in making science policy. Such a balanced performance calls for experts who not only display an informed sensitivity to the agency's mission but also enjoy unquestioned standing among their peers. Individuals who combine these traits are highly prized and may, in consequence, wield disproportionate influence. Norton Nelson's long tenure as chair of the SAB, first under William Ruckelshaus and then under Ruckelshaus's successor, Lee Thomas, can be cited as an example. Nelson's guidance undoubtedly helped to restore EPA's credibility after the trauma of the Gorsuch years and to rebuild a sense of solidarity between EPA and the scientific community. On the downside, however, such close and long-standing relations between a regulatory agency and an individual or small group of advisers may prevent the agency from receiving fresh and disinterested advice, particularly in areas of rapidly changing knowledge.[8]

Disciplinary breadth is indispensable not only for individual scientists but for committees, especially those serving EPA, and is in any event mandated to some degree by the Federal Advisory Committee Act and statutes governing specific regulatory programs. The relatively small size of advisory committees, however, is a serious stumbling block. The SAP has been criticized throughout its existence for failing to include all relevant disciplines, and the aspartame PBOI's standing was undercut by its lack of a trained toxicologist. More specific guidance from Congress about the composition of committees is unlikely to help overcome this problem. Advances in science and in the application of scientific knowledge to risk assessment make it probable that the needs of agencies will change over time, rendering detailed legislative prescriptions obsolete. For example, toxicologists and biostatisticians are now considered essential for most EPA committees, although these areas of expertise were scarcely represented in the agency's advisory network before the 1980s.

Securing a balance of political views on advisory committees presents a still more difficult problem. If one accepts the proposition that scientific knowledge is socially constructed, and that advisory committees offer a needed forum for negotiating such constructs, then it makes good conceptual sense to require that committees should include representatives of all relevant political (as well as scientific) viewpoints. Yet the idea of overt interest group representation on expert advisory committees has few adherents, even among those who concede in private that knowledgeable administrators can stack committees in favor of particular interests.[9] This reluctance finds compelling justification in the sociology of regulatory science as elaborated throughout this book. As we have seen, the authority of advisory committees in the U.S. policy system derives in part from accomplished boundary work; it is crucial for claims certified by agency advisers to be persuasively labeled "science." An openly political committee would not necessarily be able to engage in the boundary work required to bring about this result.

In sum, it appears that agencies must retain some discretion over the issue of political balance, though industry's discontent with the SAP appointment process and reports of political pressure on FDA committee appointments provide reasons to question whether uncontrolled agency discretion is desirable. Modest administrative changes might bring about the requisite level of control. For example, the removal of FDA committee appointments to higher echelons of DHHS is believed by some to have increased the opportunities for political manipulation. Here, corrective action by Congress might be helpful, though the issue requires fuller investigation than it has received in this book. In other cases, delegation of power to make nominations, if not actual appointments, to scientific bodies such as NAS or NSF might help to assure the scientific impartiality of nominees. Relying exclusively on such mechanisms, however, could perpetuate an "old boys' network" of advisers, an undesirable situation in a system where experts exercise considerable influence over policy. No matter how the power to make appointments is ultimately allocated, the problem of bias must also be countered to some extent through administrative rules dealing with such issues as vacancies, staggered terms, and conflicts of interest.

Conflicts of Interest

Many federal laws and regulations dealing with conflicts of interest explicitly address the problem of experts or advisers who have a financial stake in the matter with respect to which they are giving advice. These provisions may be more or less effective depending on how scrupulously they are enforced, but even when vigorously implemented, they conceive of conflicts of interest in overly narrow terms. A social constructivist perspective on the advisory process suggests that there are more subtle patterns of influence that should also legitimately concern policymakers.

One troubling phenomenon (noted in Chapter 7) is the appearance of former advisory committee members as consultants for one or another party to a regulatory controversy. Even when such advocacy remains strictly within the limits of the law, it may distort the review of evidence. It is only to be expected that an advisory panel will listen with special respect to one of its own former members. Recent experiences with scientific fraud and misconduct indicate that such collegiality may lead, in turn, to a suspension of critical judgment in the evaluation of scientific claims. Accordingly, it might be reasonable to place some further restrictions on agency advisers, for example, a requirement not to appear as advocates on *any* matter before a panel on which they have recently served. Recurrent appearances by the same witnesses, especially former agency officials, could also distort the dynamics of scientific advice, in the same way that repeat witnessing poses problems for courts, but this is a structural problem for which there are no easy solutions.

Targets of Review

In consulting scientific advisory bodies, agencies have to decide as a threshold matter what should be reviewed and what questions the experts should answer. Advisory processes at EPA and FDA reveal two fundamentally different approaches to these issues. FDA's advisory committees are generally asked to review the primary scientific data and to arrive at a preliminary assessment of risk. At EPA, however, the agency staff prepares the initial literature review and risk evaluation and then seeks comments from the advisory committee.

EPA's approach virtually invites divergences between the construction of data by the agency staff and that of the expert committee, which may operate, as we have seen, with very different standards of quality control and a different philosophy of risk assessment. This is precisely what happened in the cases of Alar, 2,4,5-T, nitrites, and ozone. In each instance, the agency's credibility suffered, for the elite advisory committee's judgments carried more weight than did those of the relatively less credentialed agency staff. FDA's preferred approach avoids this particular problem, for the advisers, rather than the agency, prepare the initial construction of risk. The danger for FDA is that the agency may be compelled to follow its advisers' opinion even on mixed science policy questions, though FDA officials can take comfort from the fact that they have been able to overrule their advisory committees on numerous occasions.

The Advisory Process: Procedural Options

I have argued throughout this book that submitting science policy disputes to adversarial processes promotes an unproductive deconstruction of science and fosters the appearance of capture. It follows from this analysis that the format least likely to bring about a durable consensus on contested technical issues is one that leads to confrontation between alternative constructions of uncertain scientific data. These points were highlighted in the Alar review, where EPA and Uniroyal confronted each other as disputants, while the SAP was cast in the role of an adjudicating body, a miniature science court. Deconstruction proved relentlessly one-sided in this instance, for the proceedings left no practical opportunity for EPA to contest the case presented by Uniroyal experts. Accordingly, issues that might have benefited from deeper scrutiny, such as the scientific validity of Uniroyal's bioassay criteria or the biases of company experts, were left unexamined. At the same time, SAP's judgment as to which set of claims constituted good science appeared prejudiced to the watching environmental groups and failed to win their allegiance.

Closely related to the choice of procedure is the question of who should participate in advisory committee proceedings. At a formal level, agencies have little or no discretion to insulate the process of review from public scrutiny. The requirements of federal open government laws, as well as each agency's own organic statutes, apply to almost every meeting at which agencies seek advice from a specially charged expert body. But there is also little doubt that advisory committee meetings command less attention from the environmental and consumer communities as a rule than formal administrative hearings. As revealed in dozens of meeting transcripts and interviews with agency staff, public interest groups rarely attend routine advisory committee meetings. Industry, in contrast, is generally well represented, a state of affairs that no doubt reflects the resource differences between the private sector and the public interest community.[10] Accordingly, significant policy decisions, particularly decisions not to act, may be reached after advisory deliberations that effectively engaged only one set of interests. This, in turn, may lead to skepticism about the scientific claims certified by an advisory committee.

Even when legal requirements of open meetings are met, of course, not all aspects of an agency's dealings with the scientific community are automatically open to the public. When OSTP sent its risk-assessment principles out for review, or when the panel on color additives reviewed CTFA's risk assessments, the public was not notified. Similarly, the sulfite panel first met in closed session to prepare its preliminary assessment of the literature; only when the draft document was ready to be circulated did the committee make its position known to the public. Last, but not least, experienced panelists acknowledge that there are often opportunities for private, after-hours discussions among experts even in the interstices of official public meetings. Such closed sessions not only are unavoidable but are essential to any process of scientific consensus-building that has a substantial political component.

Within these constraints, however, agencies can take steps to enhance participation. First and foremost, deliberate attempts to bypass public participation, as in the notorious case of the formaldehyde "science court," are obviously to be avoided. Second, the lay public's interests can be served in many instances through good-faith efforts to consult a broad cross-section of the expert community, so that agency conclusions are not tainted by marginal science or extremist politics. For instance, participation by scientists from federal agencies and research institutions may effectively substitute for direct involvement of citizens. Agencies can also expand the roster of potential advisers by seeking nominations through varied channels, from announcements in the *Federal Register* to solicitation of scientific professional societies. Finally, even limited opportunities for the public to participate may be adequate when science advice is not closely tied to specific regulatory decisions, as for example in the development of research policy guidelines.

A third procedural issue for agencies to consider is how often to consult a committee in the course of a single rulemaking. Repeated consultation obviously reduces the possibility that nonnegotiable differences will develop between the agency and its advisers. Most committees, however, are overworked and operate under tight time and budget constraints that make multiple reviews of the same regulatory proposal unthinkable. Exceptions can routinely be made only for such complex and economically consequential proceedings as the regular revision of NAAQS for criteria pollutants. A more modest approach that holds promise for many regulatory programs is to involve the advisory committee in looking at the evidence before the agency undertakes a comprehensive risk analysis. An early review of the adequacy of the data may prevent a scientifically unwise regulatory initiative from proceeding too far for comfortable reversal.[11] Finally, such low-cost procedural changes as EPA's practice of formally responding to the SAB chairman and committee members could foster a better understanding between regulators and their advisers and lead in the long run to more cooperative and fruitful interactions.

Scientific Advice and Judicial Review

The introduction of more numerous and, in some cases, more powerful procedures of scientific review raises the question whether judicial review has become redundant in areas of highly technical decision-making. The internal logic of the administrative process clearly provides strong reasons for courts to adopt a highly deferential posture with respect to a scientific record that has undergone thorough peer review. It is not very likely, after all, that a technically illiterate judiciary will detect flaws in scientific reasoning that has already been examined by a competent expert body. Where the court is satisfied that the agency has been made to look hard at the evidence, the standard for reversal would appropriately be strict, perhaps equivalent to that for a judgment notwithstanding a jury verdict. At the same time, courts must be careful not to defer excessively to expert judgments on matters of policy. If the record suggests that a difference of opinion between administrators and experts involved substantial policy considerations—as, for example, in the Alar decision—courts should not hesitate to probe beneath the surface of an advisory committee's recommendation and, if necessary, to overrule it.

Such is the power of scientific boundary work, however, that courts may have considerable difficulty discerning when a committee has strayed over the indeterminate border between science and policy. Other warning criteria may therefore have to be identified. In general, the arguments for close judicial scrutiny are likely to be strongest when one or more of the following conditions are met: the agency and its advisers disagree in their assessment of the evidence; the agency acts contrary to the recommendations of a peer-review panel; there is evidence of procedural impropriety in the review process. For the rest, when a smoothly functioning advisory process is in place, as in EPA's air pollution control program, the rationale for judicial intervention on substantive issues will be weak at best. In such cases the spectacle of courts immersing themselves in technical data may gradually become as much an artifact as aggressive judicial overruling of congressional enactments became in the aftermath of the New Deal.

Conclusion

The realities of scientific advice at EPA and FDA contradict many of the myths and preconceptions that have grown up around this relatively unstudied process. The notion that scientific advisers can or do limit themselves to addressing purely scientific issues, in particular, seems fundamentally misconceived. Other common myths—for example, that scientists are always conservative in assessing risks or that advice is merely a pretext for delaying decisions—also seem exaggerated. Rather, the advisory process seems increasingly important as a locus for negotiating scientific differences that carry political weight.

Scientific advice may not be a panacea for regulatory conflict or a fail-safe procedure for generating what technocrats would view as good science. It is, however, part of a necessary process of political accommodation among science, society, and the state, and it serves an invaluable function in a regulatory system that is otherwise singularly deficient in procedures for informal bargaining.

Since scientific knowledge is in perpetual flux and demands constant renegotiation, interactions involving advisory committees have to be structured in accordance with norms more flexible than those of formal and informal administrative rulemaking. Repeated rounds of analysis and review may be required before an agency reaches a conclusion that is acceptable at once to science and to the lay interests concerned with regulation. The cases discussed in Chapter 9 illustrate the futility of calling on science to cut short a policy controversy before the groundwork has been laid for accord among disparate social and political values. Adversarial procedures likewise have little to recommend them in this context, for they lead not to consensus but to counterproductive deconstructions of competing technical arguments.

Finally, the negotiated model of regulatory science suggests that the risks of science seizing the reins of decisionmaking from political institutions may have been overdrawn. Negotiation commits scientists, no less than other actors, to moderating their views toward a societal mean. The regulatory experiences of EPA and FDA indicate that it is almost inconceivable for a marginal scientific school to dominate the entire spectrum of decisions about the environment or public health and safety. The primary concern for regulators, then, is not how to guard against capture by science but how to harness the collective expertise of the scientific community so as to advance the public interest. In this effort, agencies and experts alike should renounce the naive vision of neutral advisory bodies "speaking truth to power," for in regulatory science, more even than in research science, there can be no perfect, objectively verifiable truth. The most one can hope for is a serviceable truth: a state of knowledge that satisfies tests of scientific acceptability and supports reasoned decisionmaking, but also assures those exposed to risk that their interests have not been sacrificed on the altar of an impossible scientific certainty. If advisory committees can help agencies and their adversaries to reach this kind of understanding, then the "fifth branch" will indeed have helped the fourth to come of age.

Notes

Index

Notes

1. Rationalizing Politics

1. William Lilley III and James C. Miller III, "The New 'Social' Regulation," *Public Interest* 47 (1977), pp. 49–61.

2. See, for example, Richard B. Stewart, "The Reformation of American Administrative Law," *Harvard Law Review* 88 (1975), p. 1684.

3. William Havender, "About Knowledge and Decisions," *Regulation,* March/April 1982, p. 49.

4. Keith Schneider, "Faking It: The Case against Industrial Bio-Test Laboratories," *Amicus Journal* 4 (1983), p. 14.

5. The convictions were upheld on appeal. See *United States v. Keplinger,* 776 F.2d 678 (7th Cir. 1985).

6. Harvey Brooks, "The Scientific Adviser," in Robert Gilpin and Christopher Wright, eds., *Scientists and National Policy-Making* (New York: Columbia University Press, 1964), p. 76.

7. William T. Golden, *Science and Technology Advice to the President, Congress and Judiciary* (New York: Pergamon Press, 1988). In an earlier book, Golden explicitly limited his focus to presidential science advisers: William T. Golden, *Science Advice to the President* (New York: Pergamon Press, 1980). Other works that have dealt with science advice primarily from the standpoint of national policymaking for science and technology include Gilpin and Wright, *Scientists and National Policymaking;* Daniel Kevles, *The Physicists* (New York: Alfred A. Knopf, 1971); Malcolm L. Goggin, *Governing Science and Technology in a Democracy* (Knoxville: University of Tennessee Press, 1986); Anne L. Hiskes and Richard P. Hiskes, *Science, Technology, and Policy Decisions* (Boulder, Colo.: Westview Press, 1986).

8. As the designer of the first formal proposal for a presidential science advisory committee, Golden perhaps understandably identified the White House's needs for scientific and technical advice as coextensive with those of the executive branch as a whole. The position taken in Golden's book is that the nation's policies for science entail problems that are *sui generis* and demand separate and independent analysis—a fundamentally technocratic view.

9. As an interdisciplinary field of scholarship, science policy studies include contributions from law, political science, sociology, and policy analysis. Recent

significant additions to the science policy literature include, besides works specifically mentioned in the text, the following books: Robert W. Crandall and Lester B. Lave, eds., *The Scientific Basis of Health and Safety Regulation* (Washington, D.C.: Brookings Institution, 1981); J. D. Nyhart and Milton M. Carrow, eds., *Law and Science in Collaboration* (Lexington, Mass.: Lexington Books, 1983); David M. O'Brien, *What Process Is Due?* (New York: Russell Sage Foundation, 1987).

10. Mark E. Rushefsky, *Making Cancer Policy* (New York: SUNY Press, 1986).

11. See, for example, Alice S. Whittemore, "Facts and Values in Risk Analysis for Environmental Toxicants," *Risk Analysis* 3 (1983), pp. 23–33.

12. Liora Salter, *Mandated Science* (Dordrecht, Netherlands: Kluwer, 1988).

13. These include a predominance of evaluative and screening exercises, a lack of innovation, a tendency to conservatism, and the use of inadequately established investigative methods. Salter, *Mandated Science,* pp. 187–190. Comparisons of "mandated science" controversies in different countries support Salter's conclusions about the impact of policy on the production and interpretation of science. See, for example, Sheila Jasanoff, "Cultural Aspects of Risk Assessment in Britain and the United States," in Branden B. Johnson and Vincent T. Covello, eds., *The Social and Cultural Construction of Risk* (Dordrecht, Netherlands: Reidel, 1987), pp. 359–397.

14. Joel Primack and Frank von Hippel, *Advice and Dissent* (New York: Basic Books, 1974).

15. Ted Greenwood, *Knowledge and Discretion in Government Regulation* (New York: Praeger, 1984).

16. The Administrative Procedure Act codifies the commitment to rationality through its requirement that courts should invalidate agency action that is either "arbitrary and capricious" or unsupported by "substantial evidence."

17. An early and still relevant exposition of this viewpoint may be found in Alvin M. Weinberg, "Science and Trans-Science," *Minerva* 10 (1970), pp. 209–222.

18. Authoritative support for this claim was supplied by a National Research Council committee appointed to study federal risk-management policies. See National Research Council, *Risk Assessment in the Federal Government: Managing the Process* (Washington, D.C.: National Academy Press, 1983).

19. John D. Graham, Laura C. Green, and Marc J. Roberts, *In Search of Safety: Chemicals and Cancer Risk* (Cambridge, Mass.: Harvard University Press, 1988).

20. Joseph L. Badaracco, Jr., *Loading the Dice* (Boston, Mass.: Harvard Business School Press, 1985); Ronald Brickman, Sheila Jasanoff, and Thomas Ilgen, *Controlling Chemicals: The Politics of Regulation in Europe and the United States* (Ithaca, N.Y.: Cornell University Press, 1985); Sheila Jasanoff, *Risk Management and Political Culture* (New York: Russell Sage Foundation, 1986).

21. David Collingridge and Colin Reeve, *Science Speaks to Power* (New York: St. Martin's Press, 1986), pp. 28–34.

22. Yaron Ezrahi, "Utopian and Pragmatic Rationalism: The Political Context of Scientific Advice," *Minerva* 18 (1980), p. 114.

23. William E. Akin, *Technocracy and the American Dream* (Berkeley: University of California Press, 1977). See also James Burnham, *The Managerial Revolution* (New York: John Day, 1941).

24. David Dickson, *The New Politics of Science* (New York: Pantheon, 1984).

25. Woodrow Wilson, "The Study of Administration," *Political Science Quarterly* 2 (1887), p. 201.

26. Leonard D. White, *Introduction to the Study of Public Administration* (New York: Macmillan, 1926), preface.

27. Luther Gulick, "Notes on the Theory of Organization," in Luther Gulick and Lyndall Urwick, eds., *Papers on the Science of Administration* (New York: Institute of Public Administration, 1937), pp. 3–13.

28. President's Committee on Administrative Management (Brownlow Committee), *Administrative Management in the Government of the United States* (Washington, D.C.: Government Printing Office, 1937), pp. 40–41.

29. Marver H. Bernstein, *Regulating Business by Independent Commission* (Princeton, N.J.: Princeton University Press, 1955), p. 117.

30. Don K. Price, "The Scientific Establishment," in Gilpin and Wright, *Scientists and National Policymaking,* pp. 19–20.

31. Robert C. Wood, "The Rise of an Apolitical Elite," in Gilpin and Wright, p. 43.

32. David L. Bazelon, "Risk and Responsibility," *Science* 205 (1979), pp. 277–280.

33. Derek J. de Solla Price, *Little Science, Big Science . . . and Beyond* (New York: Columbia University Press, 1986), pp. 99–100. Just three years later, however, Price observed that "in the atomic and space era, science is rapidly becoming far too important to be left to the scientists." Id., p. 138. Although Price aimed this statement at the role of scientists in making policies for science, it suggests that he was not insensitive to self-interest and other factors that might interfere with the objectivity of scientific advice in other contexts.

34. Jon D. Miller, *The American People and Science Policy* (New York: Pergamon, 1983), pp. 90–93. It should be noted that support for expert decision-makers was stronger among the attentive public than among the nonattentive. For a definition of the terms "attentive" and "nonattentive," see id., pp. 39–49.

35. Karin D. Knorr-Cetina and Michael Mulkay, eds., *Science Observed* (London: Sage, 1983); Harriet Zuckerman, "The Sociology of Science," in Neil J. Smelser, ed., *Handbook of Sociology* (Newbury Park, Calif.: Sage, 1988), ch. 6.

36. See, in this connection, Bruno Latour and Steve Woolgar, *Laboratory Life* (Princeton, N.J.: Princeton University Press, 1986). Viewing a laboratory from the vantage point of an anthropological observer, Latour and Woolgar documented how a single scientific "fact"—the structure of thyrotropin-releasing hormone (TRH)—was "freed from the circumstances of its production" (p. 105) and entered the body of accepted scientific knowledge. Their book provides an absorbing account of the delicate negotiations ("microprocessing") that take place within the laboratory in order to convert a body of subjective observations and statements into a "fact."

37. On the social construction of risk, see Mary Douglas and Aaron Wildavsky, *Risk and Culture* (Berkeley: University of California Press, 1982); Jasanoff,

Risk Management and Political Culture; Johnson and Covello, eds., *The Social and Cultural Construction of Risk*. For a useful introduction to the literature on policy controversies involving science and technology, see the case studies compiled in Dorothy Nelkin, ed., *Controversy,* 2d ed. (Beverly Hills, Calif.: Sage Publications, 1984).

38. Latour and Woolgar, *Laboratory Life,* p. 179.

39. I have described this phenomenon in greater detail in two earlier papers. See Sheila Jasanoff, "Contested Boundaries in Policy-Relevant Science," *Social Studies of Science* 17 (1987), pp. 195–230, and "The Problem of Rationality in American Health and Safety Regulation," in Roger Smith and Brian Wynne, eds., *Expert Evidence: Interpreting Science in the Law* (London: Routledge, 1989).

40. Thomas S. Kuhn, *The Structure of Scientific Revolutions,* 2d ed. (Chicago: University of Chicago Press, 1970).

41. Harry M. Collins, *Changing Order* (London: Sage, 1985), pp. 83–89.

42. Id., pp. 87–88.

43. Thomas F. Gieryn, "Boundary-Work and the Demarcation of Science from Non-Science: Strains and Interests in Professional Ideologies of Scientists," *American Sociological Review* 48 (1983), pp. 781–795.

44. David Dickson has characterized the technocratic approach as favoring "solutions reached through a consensus of experts." *The New Politics of Science,* p. 219. His book traces a historical oscillation between periods of high and low scientific autonomy or technocracy and democracy in the interactions between scientists and governmental decisionmakers.

45. Don K. Price, *The Scientific Estate* (Cambridge, Mass.: Harvard University Press, 1965), pp. 121–122.

2. Flawed Decisions

1. Edith Efron, *The Apocalyptics* (New York: Simon and Schuster, 1984).

2. The origin of the environmental movement in America is conventionally dated to the publication of Carson's *Silent Spring* in 1962. While Carson was troubled primarily by the environmental effects of chemicals, concerns about human health formed the focal point of Samuel S. Epstein's extremely influential book on environmental carcinogens, *The Politics of Cancer* (New York: Sierra Club Books, 1978).

3. Incorporated into the 1958 amendment to the Food, Drug, and Cosmetic Act, the Delaney clause provides that no food additive may be cleared for use if it is found, on the basis of animal or human evidence, to induce cancer. 21 U.S.C. §348(c)(3)(A).

4. For an articulate exposition of these views, see Nicholas A. Ashford, C. William Ryan, and Charles C. Caldart, "Law and Science Policy in Federal Regulation of Formaldehyde," *Science* 222 (1983), pp. 894–900.

5. For contrasting responses to these questions, see the letters of Devra L. Davis and Cyril Comar in *Science* 203 (1979), p. 7.

6. In his announcement of the proposed saccharin ban, FDA's Acting Commissioner Sherwin Gardner compared the dosage given to rats in a Canadian study to drinking 800 12-ounce cans of diet soda a day. Reginald W. Rhein, Jr.,

and Larry Marion, *The Saccharin Controversy* (New York: Monarch Press, 1977), pp. 4–5.

7. Saccharin Study and Labeling Act, Pub. L. 95–203, 95th Congress, 1st Session, 1977.

8. R. Jeffrey Smith, "Ever So Cautiously, the FDA Moves toward a Ban on Nitrites," *Science* 201 (1978), pp. 887–891.

9. Id.

10. Id.

11. R. Jeffrey Smith, "Approval Sought for Nitrite Plan," *Science* 204 (1979), p. 386.

12. Id.

13. R. Jeffrey Smith, "Nitrites: FDA Beats a Surprising Retreat," *Science* 209 (1980), pp. 1100–1101.

14. Id.

15. Id.

16. This phenomenon is discussed in detail in William Broad and Nicholas Wade, *Betrayers of the Truth* (Oxford: Oxford University Press, 1985), pp. 107–125.

17. Smith, "Nitrites: FDA Beats a Surprising Retreat."

18. Interview with Paul Newberne, Ithaca, N.Y., February 2, 1981.

19. Interview with Donald Kennedy, Ithaca, N.Y., October 1980.

20. Smith, "Nitrites: FDA Beats a Surprising Retreat."

21. Thomas Whiteside, *The Pendulum and the Toxic Cloud: The Course of Dioxin Contamination* (New Haven: Yale University Press, 1979).

22. R. Jeffrey Smith, "EPA Halts Most Uses of Herbicide 2,4,5-T," *Science* 203 (1979), pp. 1090–1091.

23. Id.

24. R. Jeffrey Smith, "Court Reluctantly Upholds EPA on 2,4,5-T Suspension," *Science* 204 (1979), p. 602.

25. Transcript of 25th Meeting of the FIFRA Scientific Advisory Panel, vol. I, August 15, 1979, pp. 204–205 (statement of Dr. Edward Smuckler). Interestingly, Smuckler later represented other chemical producers before the SAP.

26. The absence of such data led Britain's Advisory Committee on Pesticides to reject the Alsea study as a valid basis for assessing the risk of human exposure to 2,4,5-T. See Advisory Committee on Pesticides, *Further Review of the Safety of the Herbicide 2,4,5-T* (London: HMSO, 1980).

27. The case was finally settled out of court in 1984 and the settlement agreement was upheld by the U.S. Supreme Court in 1988. An extremely readable account of the proceedings leading to the settlement may be found in Peter Schuck, *Agent Orange on Trial* (Cambridge, Mass.: Harvard University Press, 1986).

28. Ultimately, the federal government picked up 90 percent of the cost ($33 million) and the state government the remaining $3.7 million. Michael Gough, *Dioxin, Agent Orange* (New York: Plenum Press, 1986), p. 134.

29. "Dow Ends Fight over 2,4,5-T Safety," *Chemical and Engineering News*, October 24, 1983, p. 5.

30. Gina Kolata, "Love Canal: False Alarm Caused by Botched Study," *Science* 208 (1980), pp. 1239–1242.

31. Letter of Dante Picciano, *Science* 209 (1980), pp. 754, 756.

32. Kolata, "Love Canal."

33. Id.

34. Id., p. 1251. See also Barbara J. Culliton, "Continuing Confusion over Love Canal," *Science* 209 (1980), pp. 1002–1003.

35. Report of the Governor's Panel to Review Scientific Studies and the Development of Public Policy on Problems Resulting from Hazardous Wastes, New York, October 8, 1980.

36. Letter of Stephen Gage, *Science* 209 (1980), pp. 752, 754.

37. Letter of Margery Shaw, *Science* 209 (1980), pp. 751–752.

38. Letter of Dante Picciano, *Science* 209.

39. Kolata, "Love Canal."

40. Id. Low rates in the controls would, of course, increase the likelihood of finding positive effects in the study population.

41. Letter of Dante Picciano, *Science* 209.

42. Id.

43. Letter of Margery Shaw, *Science* 209.

44. NCI-NIEHS-NIOSH, "Estimates of the Fraction of Cancer in the U.S. Related to Occupational Factors" (manuscript, Washington, D.C., 1978).

45. OSHA, "Identification, Classification and Regulation of Potential Occupational Carcinogens," *Federal Register* 45 (January 22, 1980), pp. 5001–5296.

46. R. Jeffrey Smith, "Government Says Cancer Rate Is Increasing," *Science* 209 (1980), pp. 998–1000, 1002.

47. Id.

48. Richard Peto, "Distorting the Epidemiology of Cancer: The Need for a More Balanced Overview," *Nature* 284 (1980), pp. 297–300.

49. Efron, *The Apocalyptics*.

50. Peto, "Distorting the Epidemiology of Cancer," p. 300.

51. Richard Doll and Richard Peto, "The Causes of Cancer: Quantitative Estimates of Avoidable Risks of Cancer in the United States Today," *Journal of the National Cancer Institute* 66 (1981), pp. 1193–1308.

52. National Research Council, *Risk Assessment in the Federal Government*.

53. American Industrial Health Council, "AIHC Recommended Framework for Identifying Carcinogens and Regulating Them in Manufacturing Situations," October 11, 1979; "AIHC Proposal for a Scientific Panel" (mimeo), March 12, 1981. In order to attract the "best" scientists, AIHC recommended that appointments to the panel be part time and for a period of three years. Public participation was to be invited through informal notice and comment procedures.

54. National Science Council Act, H.R. 638 (Wampler Bill), 97th Congress, 1st Session (1981).

55. National Research Council, *Risk Assessment in the Federal Government*.

56. Id., p. 142.

57. Id.

58. Id., p. 5.

59. Id., pp. 156–60.

60. These proposals provided a starting point for a more detailed analysis of peer-review options carried out for EPA by the American Chemical Society (ACS)

and the Conservation Foundation (CF), a Washington-based policy research organization. The project formally defined peer review in the policy context as "the critical scrutiny of a policy statement by independent technical experts to determine 1) the accuracy of technical data, 2) the validity of technical interpretations, and 3) the relevance of the technical data and interpretations to a policy decision. Peer review is used at various stages of the policymaking process to reduce the chance of omission or mistaken application of key technical material." The ACS-CF report also produced a "road map" for the development of peer-review guidelines at EPA; in other words, a description of options available to the agency and a discussion of their strengths, weaknesses, and limitations. Given its modest objectives, it is perhaps not surprising that the report did not differentiate between genuinely difficult questions, such as the problem of controlling bias, and more minor questions of management, such as the appropriate fees and terms of appointment for panel members. American Chemical Society and Conservation Foundation, "Issues in Peer Review of the Scientific Basis for Regulatory Decisions" (Washington, D.C.: American Chemical Society, 1985).

61. 42 U.S.C. 4365 (1978).

62. Clean Air Act §109(d)(2). CASAC functions formally as a subcommittee of the Science Advisory Board.

63. FIFRA §25(d).

64. FDCA §513(b), (c).

65. See, for example, Food Safety and Modernization Act of 1983, S.1938, *Congressional Record,* October 6, 1983, S 13780.

66. Nicholas A. Ashford, "Advisory Committees in OSHA and EPA: Their Use in Regulatory Decisionmaking," *Science, Technology, and Human Values* 9 (1984), p. 75.

67. Specifically, review has to take place before the agency issues a notice of proposed rulemaking. CPSA §28.

68. CERCLA Amendment §110.

3. Science for the People

1. There is an extensive interdisciplinary literature on these topics. See, for example, William Lilley and James C. Miller, "The New 'Social' Regulation," *Public Interest* 47 (1977), pp. 49–61; Eugene Bardach and Robert A. Kazan, *Social Regulation, Strategies for Reform* (San Francisco: Institute for Contemporary Studies, 1978); Barry Mitnick, *The Political Economy of Regulation: Creating, Designing and Removing Regulatory Forms* (New York: Columbia University Press, 1980); James Q. Wilson, *The Politics of Regulation* (New York: Basic Books, 1980); Michael S. Baram, *Alternatives to Regulation* (Lexington, Mass.: Lexington Books, 1982); R. Shep Melnick, "Pollution Deadlines and the Coalition for Failure," *Public Interest* 75 (1984), pp. 123–134; idem, *Regulation and the Courts* (Washington, D.C.: Brookings Institution, 1983).

2. I am using the term "public" here both in the sense of "belonging to government" and in the sense of "belonging to the people." Since the early 1970s, governmental agencies as well as the lay public have played a much more active

role than at any previous period in the production, interpretation, and certification of scientific knowledge related to policy.

3. National Research Council, *Toxicity Testing* (Washington, D.C.: National Academy Press, 1982).

4. A number of statutes implicitly delegate to the regulatory agencies a responsibility to check the quality of the data supporting governmental intervention. An example is the Occupational Safety and Health Act (OSH Act), which provides that the "best available evidence" should be used in making decisions about toxic substances in the workplace. 29 U.S. Code §655 (b). The "best available" standard implicitly contains a mandate to assess the quality of the evidence. Similarly, the Delaney clause of the Food, Drug, and Cosmetic Act authorizes FDA to determine whether studies used to evaluate the carcinogenicity of food additives are "appropriate" for the purpose. 21 U.S. Code §348(c)(3)(A). Evaluating the quality of the relevant studies is clearly integral to the task of establishing their "appropriateness."

5. 42 U.S. Code §1857 et seq.

6. 15 U.S. Code §2601 et seq.

7. See, for example, Marshall S. Shapo, *A Nation of Guinea Pigs* (New York: Free Press, 1979), p. 154.

8. 7 U.S. Code §136 et seq.

9. For an extended discussion of these balancing provisions in federal health and safety statutes, see Michael S. Baram, "Cost-Benefit Analysis: An Inadequate Basis for Health, Safety, and Environmental Regulatory Decisionmaking," *Ecology Law Quarterly* 8 (1980), pp. 409–472.

10. Institute of Medicine, *Responding to Health Needs and Scientific Opportunity: The Organizational Structure of the National Institutes of Health* (Washington, D.C.: National Academy Press, 1984).

11. NTP, Fiscal Year 1985 Annual Plan (Washington, D.C.: DHHS, 1985), p. 13.

12. For a brief history of the growth in EPA's budget and personnel resources between 1970 and 1981, and the agency's subsequent loss of strength under President Reagan, see Walter A. Rosenbaum, *Environmental Politics and Policy* (Washington, D.C.: Congressional Quarterly, 1985).

13. GAO, *Attrition of Scientists and Engineers at Seven Agencies* (Washington, D.C.: GAO, 1984), p. 6.

14. EPA, *Long-Range Research Agenda for the Period 1986–1991* (Washington, D.C., April 1985). See also *Chemical and Engineering News*, December 1985, p. 12.

15. EPA, *Long-Range Research Agenda*, p. 3.

16. GAO, *Attrition of Scientists*, p. 11.

17. Charles Marwick, "FDA Center Emphasizes Research," *The Scientist*, December 14, 1987, p. 7.

18. Ronald W. Hart and Arthur R. Norris, "Managing the Government's Toxicological Research Center," *Concepts of Toxicology*, vol. 1 (Basel: Karger, 1984), pp. 258–259.

19. The good laboratory practices (GLP) regulations promulgated by FDA in

December 1978 responded to the discovery of defects in a large number of safety studies submitted to FDA by industry. The GLP regulations are intended to assure the quality and integrity of nonclinical laboratory studies carried out in support of research or marketing permits for a variety of products regulated by the agency.

20. National Cancer Program, 1983–1984 Director's Report and Annual Plan, FY 1986–1990 (Washington, D.C.: DHHS, 1984), p. 35.

21. See Note, "The Federal Advisory Committee Act," *Harvard Journal on Legislation* 10 (1973), p. 217.

22. *Nader v. Baroody,* 396 F.Supp. 1231 (D.D.C. 1975).

23. *Lombardo v. Handler,* 397 F.Supp. at 797–800; Jerry W. Markham, "The Federal Advisory Committee Act," *University of Pittsburgh Law Review* 35 (1974), p. 574.

24. However, one court has invoked FACA to enjoin a proposed closed work session on criteria documents for air quality standards involving scientists who were invited by EPA and paid to participate as consultants. *American Iron and Steel Institute v. Costle,* 12 Environment Reporter Cases 1271 (W.D.Pa. 1979); see also 10 Environment Reporter (BNA) 196 (1979).

25. 5 U.S.C. App. I, §§5(b)(2), 5(c).

26. *National Anti-Hunger Coalition v. Executive Committee,* 557 F.Supp. 524, 528 (D.D.C. 1983)(committee composed solely of executives of major corporations did not violate FACA where committee's function was "narrow and explicit" and its purpose was "to apply to federal programs the expertise of leaders in the private sector"); see also *Harvard Journal on Legislation* 10 (1973), p. 229. But for a suggestion that this reading is too narrow, see Markham, "Federal Advisory Committee Act," p. 588.

27. Some writers have argued that science and policy are so closely intertwined in most discussions of regulatory science that FACA's fair-balance concept should always be broadly interpreted. See Nicholas A. Ashford, "Advisory Committees in OSHA and EPA: Their Use in Regulatory Decisionmaking," *Science, Technology, and Human Values* 9 (1984), pp. 73–74. This reading of the law would be more burdensome for agencies, since they would be called upon to ensure adequate representation along a number of disparate dimensions: disciplinary background, institutional affiliation, and political allegiance. Such a practice might also expose agencies to charges that they were politicizing science (see Chapter 5).

28. *Metcalf v. National Petroleum Council,* 553 F.2d 176, 186 (D.C. Cir. 1977).

29. See, for example, *Center for Auto Safety v. Tiemann,* 414 F.Supp. 215, 220 (D.D.C. 1976), remanded on other grounds, 580 F.2d 689 (D.C. Cir. 1978); *Nader v. Baroody,* 396 F.Supp. 1231–32 (D.D.C. 1975); *Consumers Union v. HEW,* 409 F.Supp. 472, 475 (D.D.C. 1975).

30. *Physicians' Education Network v. HEW,* 653 F.2d 621, 623 (D.C. Cir. 1981).

31. *National Anti-Hunger Coalition v. Executive Committee,* 711 F.2d 1071, 1074, n.2 (D.C. Cir. 1983).

32. *American Iron and Steel Institute v. Costle.*

33. Jane Lang McGrew, "How to Let in the Sunshine without Getting Burned: Protecting Your Rights before Advisory Committees," *Food Drug Cosmetic Law Journal* 30 (1975), p. 536.

34. This indeed was the position taken by the court in *American Petroleum Institute v. Costle,* 665 F.2d 1176 (D.C. Cir. 1981) (see below).

35. 5 U.S.C. §552(b)(5).

36. Jeffrey A. Sar, "The Federal Advisory Committee Act: A Key to Washington's Back Door," *South Dakota Law Review* 20 (1975), pp. 391–399. Note, for example, the use of the OMB guidelines on January 9, 1975, to close a meeting of the Panel on Revision of Laxatives, Anti-diarrheal, Anti-emetic and Emetic Drugs (id., p. 399).

37. *Wolfe v. Weinberger,* 403 F.Supp. 238 (D.D.C. 1975).

38. 5 U.S.C. §552(b)(4).

39. The characteristics of science policy have been most exhaustively explored in connection with federal policies for determining carcinogenic risk and regulating suspected carcinogens. For an exceptionally competent treatment of the subject from a lawyer's perspective, see Thomas O. McGarity, "Substantive and Procedural Discretion in Administrative Resolution of Science Policy Questions: Regulating Carcinogens in EPA and OSHA," *Georgetown Law Review* 67 (1979), pp. 729–810.

40. *Ethyl Corp. v. EPA,* 541 F.2d 1 (D.C. Cir. 1976); *Reserve Mining Co. v. EPA,* 514 F.2d 492 (8th Cir. 1975); *Certified Color Manufacturers Association v. Mathews,* 543 F.2d 284 (D.C. Cir. 1976); *EDF v. EPA,* 598 F.2d 62 (D.C. Cir. 1978); *Society of Plastics Industry, Inc. v. OSHA,* 509 F.2d 1301 (2nd Cir. 1975).

41. *EDF v. EPA,* 489 F.2d 1247 (D.C. Cir. 1973); *EDF v. EPA* 548 F.2d 998 (D.C. Cir. 1976); *EDF v. EPA,* 598 F.2d 62 (D.C. Cir. 1978).

42. *Ethyl Corp. v. EPA; EDF v. EPA,* 510 F.2d 1292 (D.C. Cir. 1975); *Hercules Inc. v. EPA,* 598 F.2d 91 (D.C. Cir. 1978); *Lead Industries Association v. EPA,* 647 F.2d 1130 (D.C. Cir. 1980); *United Steelworkers of America v. Marshall,* 647 F.2d 1189 (D.C. Cir. 1980).

43. *Ethyl Corp. v. EPA,* p. 28.

44. Id., n. 58. Current opinion in the legal community continues to hold that juries have wide latitude in determining facts on the basis of contested evidence. For a recent exposition of these views, see Michael J. Saks, "Accuracy v. Advocacy," *Technology Review,* August/September 1987, pp. 43–49.

45. *EDF v. EPA,* 465 F.2d 528 (D.C. Cir. 1972), p. 538.

46. *Certified Color Manufacturers Association,* pp. 297–298.

47. 469 F.Supp. 892 (E.D. Mich. 1979).

48. 598 F.2d 91 (D.C. Cir. 1978), p. 126.

49. *Seacoast Anti-Pollution League v. Costle,* 572 F.2d 872 (1st Cir. 1978).

50. See note 41.

51. 598 F.2d 91, pp. 107–8.

52. *International Harvester Co. v. Ruckelshaus,* 478 F.2d 615 (D.C. Cir. 1973).

53. Id., p. 649.

54. William F. Pedersen, Jr., "What Judges Should Know about Risk," *Natural Resources and Environment* 1 (1986), p. 38.

55. This standard is applied to "informal rulemaking" under the APA. This procedure requires agencies to give notice of their proposed actions, receive comments from the public, and announce their final decisions accompanied by a "concise general statement of their basis and purpose." 5 U.S.C. §706(2)(A).

56. This standard is associated with "formal adjudication," a procedure that resembles a trial and calls for procedural safeguards similar to those used in court. 5 U.S.C. §706(2)(E).

57. See, for example, *Lead Industries Association v. EPA,* 647 F.2d 1130, 1146 n.2 (D.C. Cir. 1980); *Pacific Legal Foundation v. Department of Transportation,* 593 F.2d 1338, 1343 n.35 (D.C. Cir. 1979); *Associated Industries v. Department of Labor,* 487 F.2d 342, 350 (2d Cir. 1973). The Supreme Court defined "substantial evidence" as merely "such evidence as a reasonable mind might accept as adequate to support a conclusion." *Consolidated Edison Co. v. National Labor Relations Board,* 305 U.S. 197 (1938), p. 229.

58. Antonin Scalia and Frank Goodman, "Procedural Aspects of the Consumer Product Safety Act," *UCLA Law Review* 20 (1973), p. 935.

59. In interpreting statutes, courts generally defer to the implementing agency's construction of the law so long as it is "reasonable and not in conflict with the expressed intent of Congress." *United States v. Riverside Bayview Homes, Inc.,* 106 S. Ct. 455, 461 (1985), citing *Chemical Manufacturers Association v. Natural Resources Defense Council, Inc.,* 470 U.S. 116 (1985); and *Chevron U.S.A., Inc. v. Natural Resources Defense Council, Inc.,* 467 U.S. 837 (1984). Procedural review is limited to an agency's compliance with statutory procedural requirements and any other procedures voluntarily adopted by the agency. Courts may not impose any additional procedures on an agency. See *Vermont Yankee Nuclear Power Corp. v. Natural Resources Defense Council, Inc.,* 435 U.S. 519 (1978), pp. 542–549.

60. Harold Leventhal, "Environmental Decisionmaking and the Role of the Courts," *University of Pennsylvania Law Review* 122 (1974), p. 514. See also *Citizens to Preserve Overton Park v. Volpe,* 401 U.S. 402 (1971); William H. Rodgers, Jr., "A Hard Look at *Vermont Yankee:* Environmental Law under Close Scrutiny," *Georgetown Law Review* 67 (1979), pp. 699–727.

61. *Weyerhauser Co. v. Costle,* 590 F.2d 1011 (D.C. Cir. 1978), p. 1027.

62. Thomas O. McGarity, "Beyond the Hard Look: A New Standard for Judicial Review," *Natural Resources and Environment* 2 (1986), pp. 32–34, 66–68.

63. The D.C. Circuit held in 1971 that certain adjudicatory procedures were required even in informal rulemaking. *Walter Holm and Co. v. Hardin,* 449 F.2d 1009 (D.C. Cir. 1971). The *International Harvester* case, decided two years later, proved even more influential as a rationale for imposing formal constraints on the consideration of regulatory science. The court suggested in this case that a right of cross-examination might be judicially recognized where less formal procedures "proved inadequate to probe 'soft' and sensitive subjects and witnesses." 478 F.2d 615, 631 (D.C. Cir. 1973). Another case decided in the same year set aside an agency decision for failure to include adjudicatory procedures. *Mobil Oil Corp. v. Federal Power Commission,* 483 F.2d 1238 (D.C. Cir. 1973); but see *Wisconsin Gas V. FERC,* 770 F.2d 1144, 1167 (D.C. Cir. 1985)(declaring that *Mobil Oil* is

no longer good law). As in *International Harvester,* the *Mobil Oil* court seemed concerned that factual disputes had not been aired thoroughly enough to create the kind of record required for effective judicial review. Seeing which way the judicial winds were blowing, the agencies began voluntarily opting for more formal procedures.

64. For one statement of this kind, see *Ethyl Corp. v. EPA,* p. 67 (concurring opinion of Judge Bazelon).

65. David Bazelon, "Coping with Technology through the Legal Process," *Cornell Law Review* 62 (1977), p. 823.

66. See, for example, concurring opinion of Judge Bazelon in *International Harvester,* p. 652.

67. *Natural Resources Defense Council v. U.S. Nuclear Regulatory Commission,* 547 F.2d 633 (D.C. Cir. 1976).

68. Id., p. 653.

69. Bazelon evidently assumed that debate among technical adversaries would create a record in which uncertainties would be fully disclosed and rigorously probed. Id.

70. For a general discussion of these issues, see Ronald Brickman, Sheila Jasanoff, and Thomas Ilgen, *Controlling Chemicals: The Politics of Regulation in Europe and the United States* (Ithaca, N.Y.: Cornell University Press, 1985). See also Sheila Jasanoff, *Risk Management and Political Culture* (New York: Russell Sage Foundation, 1986).

71. In such proceedings, representatives from government, industry, and environmental groups attempt to resolve their disagreements with the assistance of a neutral mediator. Negotiated rulemaking is viewed as an effective means of resolving not only political differences but also disputes over science. Philip Harter, "Negotiating Regulations: A Cure for the Malaise," *Environmental Impact Assessment Review* 3 (1982), pp. 75–92.

72. 435 U.S. 519 (1978).

73. Justices Blackmun and Powell did not participate in this case.

74. *Vermont Yankee,* pp. 547–548.

75. 462 U.S. 87 (1983).

76. Id., p. 103.

77. 657 F.2d 298 (D.C. Cir. 1981).

78. Id., pp. 398–399.

79. Id., pp. 400–401. In reaching these conclusions, the court was clearly swayed by *Vermont Yankee*'s message that judges should hesitate to impose their notions of procedural propriety upon a regulatory agency seeking to carry out a congressional mandate.

80. Id., p. 408.

81. Id., p. 409.

82. 448 U.S. 607 (1980).

83. Id., p. 645. But see id., p. 655, where the court cautioned that the determination of significant risk should not be viewed as a "mathematical straitjacket."

84. OSHA justified this determination on policy grounds, claiming that risk assessment was too new and uncertain a technique to use as a basis for regulatory

policy. The Supreme Court's decision has been criticized as a return to judicial interventionism, transgressing the subtle divide between legitimate control of administrative decisionmaking and impermissible judicial lawmaking. See, for example, Devra Davis, "The 'Shotgun Wedding' of Science and Law: Risk Assessment and Judicial Review," *Columbia Journal of Environmental Law* 10 (1985), pp. 78–82. See also McGarity, "Beyond the Hard Look."

85. 701 F.2d 1137 (5th Cir. 1983).

86. Id., p. 1145.

87. Davis, "Shotgun Wedding." See also Nicholas A. Ashford, C. William Ryan, and Charles C. Caldart, "Law and Science Policy in Federal Regulation of Formaldehyde," *Science* 222 (1983), pp. 894–900; Kenneth S. Abraham and Richard A. Merrill, "Scientific Uncertainty in the Courts," *Issues in Science and Technology* 2 (1986), pp. 93–107.

88. The impact of this case on EPA decisionmaking is discussed at greater length in Chapter 6.

89. Although EPA submitted the draft criteria document to the SAB, the Board never saw either the final criteria document or the final standard. The court held that EPA had satisfied the legal duty to consult with respect to the criteria document, but not with respect to the standard. However, the court did not consider this error sufficiently serious to invalidate the standard.

90. For a more extended development of this argument, see Sheila Jasanoff, "Contested Boundaries in Policy-Relevant Science," *Social Studies of Science* 17 (1987), pp. 195–230.

4. Peer Review and Regulatory Science

1. See, for example, Stephen Lock, *A Difficult Balance: Editorial Peer Review in Medicine* (London: Nuffield Provisional Hospitals Trust, 1985), p. 4.

2. Robert K. Merton, "The Normative Structure of Science," rpt. in Merton, *The Sociology of Science* (Chicago: University of Chicago Press, 1973), p. 268.

3. See, particularly, Michael J. Mulkay, "Norms and Ideology in Science," *Social Science Information* 15 (1976), pp. 637–656, and "Interpretation and the Use of Rules: The Case of the Norms of Science" in Thomas F. Gieryn, ed., *Science and Social Structure* (New York: New York Academy of Sciences, 1980), pp. 172–196.

4. The conventional dating of scientific refereeing to this event has been provocatively described by Arie Rip as the "founding myth" of peer review. He asks "why the Jesuit astronomers of the Collegium Romanum, who critically evaluated Galileo's claims, and the mathematicians who advised about the imprimatur of his books, are never mentioned as the fathers of peer review." See Arie Rip, "Commentary: Peer Review Is Alive and Well in the United States," *Science, Technology, and Human Values* 10 (1985), p. 84.

5. Robert K. Merton and Harriet Zuckerman, "Institutionalized Patterns of Evaluation in Science," rpt. in Merton, *The Sociology of Science,* p. 469.

6. Sheila Jasanoff, "Peer Review in the Regulatory Process," *Science, Technology, and Human Values* 10 (1985), p. 22.

7. About 75 percent of major scientific journals currently use some form of

external peer review, and this fraction may well increase over time. Lock, *A Difficult Balance,* p. 3.

8. Id., pp. 6–7.

9. The most important exception to peer review in the federal grants process in recent years has been the direct funding of research projects by Congress at specified academic institutions. For an account of this type of "pork barrel politics," see Colin Norman and Eliot Marshall, "Over a (Pork) Barrel: The Senate Rejects Peer Review," *Science* 233 (1986), pp. 145–146.

10. Institute of Medicine, *Responding to Health Needs and Scientific Opportunity: The Organizational Structure of the National Institutes of Health* (Washington, D.C.: National Academy Press, 1984).

11. Eliot Marshall, "Peer Review Comes to ADAMHA," *Science* 204 (1979), pp. 601–602.

12. Letter of Stephen J. Gage, *Science* 209 (1980), p. 752.

13. Interview with Marcia Angell, Deputy Editor, *New England Journal of Medicine,* Boston, November 14, 1985.

14. Interview with Stephen Lock, London, June 1986.

15. Philip M. Boffey, "Hope Seen for Limited Breast Cancer Surgery," *New York Times,* March 14, 1985, p. A1.

16. *Allen v. United States,* 588 F.Supp. 247 (D.D. Utah 1984).

17. "Journal Publishes Disputed Study on Cancer Increase in Mormons," *New York Times,* January 13, 1984.

18. Charles Perrow, *Normal Accidents* (New York: Basic Books, 1984).

19. See, for example, Lock, *A Difficult Balance,* p. 3.

20. Id., p. 22.

21. Douglas P. Peters and Stephen J. Ceci, "Peer Review Practices of Psychological Journals: The Fate of Published Articles Submitted Again," *Behavioral and Brain Sciences* 5 (1982), pp. 187–195.

22. Michael J. Mahoney, "Publication Prejudices: An Experimental Study of Confirmatory Bias in the Peer Review System," *Cognitive Therapy and Research* 1 (1977), pp. 161–175.

23. For example, Karl Meyer, then an unknown postgraduate student, was unable to publish his original paper on the first law of thermodynamics in *Annales der Physik* in 1842. See Lock, *A Difficult Balance,* p. 27. Gregor Mendel's pathbreaking work on genetics was rejected by C. Nageli and remained unknown for a generation, possibly because Mendel, "a Catholic priest in far away Moravia," could not command high credibility with a mainstream biologist and intellectual competitor. See Ernst Mayr, *The Growth of Biological Thought* (Cambridge, Mass.: Harvard University Press, 1982), p. 723.

24. Robert K. Merton, "The Matthew Effect in Science," in Merton, *The Sociology of Science,* pp. 439–459.

25. Harriet Zuckerman, *Scientific Elite: Nobel Laureates in the United States* (New York: Free Press, 1977).

26. Mahoney, "Publication Prejudices."

27. Lock, *A Difficult Balance,* p. 33.

28. Stephen Cole, Jonathan Cole, and Gary A. Simon, "Chance and Consensus in Peer Review," *Science* 214 (1981), pp. 881–886; Jonathan Cole and

Stephen Cole, "Which Researcher Will Get the Grant," *Nature* 279 (1979), pp. 575–576.

29. Cole, Cole, and Simon, "Chance and Consensus," p. 885.

30. Lock, *A Difficult Balance,* p. 7.

31. Some examples of this type of mistake are cited in Walter W. Stewart and Ned Feder, "The Integrity of the Scientific Literature," *Nature* 325 (1987), pp. 207–214.

32. Journalistic support for this claim may be found in Malcolm W. Browne, "Fusion in a Jar: Announcement by 2 Chemists Ignites Uproar," *New York Times,* March 28, 1989, p. C1.

33. A full account of Alsabti's career is provided in William Broad and Nicholas Wade, *Betrayers of the Truth* (Oxford: Oxford University Press, 1985), pp. 38–55.

34. Lawrence K. Altman, "Eminent Harvard Professor Quits over Plagiarism, University Says," *New York Times,* November 29, 1988, pp. A1, A22.

35. Broad and Wade, *Betrayers,* pp. 203–211.

36. Id., pp. 205–206.

37. Leon J. Kamin, *The Science and Politics of IQ* (Potomac, Md.: Lawrence Erlbaum, 1974).

38. Leslie S. Hearnshaw, *Cyril Burt, Psychologist* (Ithaca, N.Y.: Cornell University Press, 1979).

39. Broad and Wade, *Betrayers,* p. 210.

40. Id., pp. 89–96.

41. *Fraud in Biomedical Research,* Hearings before the Subcommittee on Investigations and Oversight, Committee on Science and Technology, U.S. House of Representatives, 97th Congress, March 31-April 1, 1981, pp. 65–66.

42. Broad and Wade, *Betrayers,* p. 71.

43. Id., pp. 66–67. Similarly, two non-Cornell scientists who were also unable to repeat Spector's experiments, but who did not have access to his raw data, did not challenge the validity of his work or his integrity as a researcher. Id., pp. 70–71.

44. Id., pp. 13–15.

45. Stewart and Feder, "Integrity."

46. See Philip M. Boffey, "Major Study Points to Faulty Research at Two Universities," *New York Times,* April 22, 1986, pp. C1, C11.

47. In an interview with this author, Lewin expressed even stronger reservations about the whole concept of historical controls in biomedical research. Calling attention to the complexity of biological systems, he suggested that the use of historical controls would introduce an unacceptable degree of uncontrolled variation into any experiment. Interview with Benjamin Lewin, Hedgesville, W.V., September 1987.

48. Stewart and Feder, "Integrity," pp. 207–214.

49. Eugene Braunwald, "On Analysing Scientific Fraud," *Nature* 325 (1987), p. 216.

50. "Fraud, Libel and the Literature," *Nature* 325 (1987), p. 182.

51. The term "regulatory science" has gained greater currency in the United States than the roughly synonymous term "mandated science" used by Liora

Salter. I accordingly use the former term in this chapter and throughout the book, although my discussion draws both on work by Rushefsky and other U.S. writers on regulatory science and on Salter's work on mandated science. There is similarly no unanimity among science policy analysts about the best way to refer to science unrelated to policy. "Normal science," the term favored by Rushefsky, has been used in another and better-known sense by Kuhn in his work on scientific revolutions. For my purposes, "research science" seems the most desirable formulation.

52. Barry Barnes and David Edge, *Science in Context* (Milton Keynes, U.K.: Open University Press, 1982), p. 147.

53. Id.

54. Liora Salter, *Mandated Science* (Dordrecht, Netherlands: Kluwer, 1988), pp. 187–188.

55. Alvin M. Weinberg, "Science and Its Limits: The Regulator's Dilemma," *Issues in Science and Technology*, Fall 1985, p. 68. The term "regulatory science" in this context seems at least partly coextensive with the concept of "trans-science" used by Weinberg in an earlier article, "Science and Trans-Science," *Minerva* 10 (1972), pp. 209–222.

56. The sociologist Bernard Barber has noted, for example, that "the chief dwelling of 'pure' science is in the university and of 'applied' science in government and industry." Barber, *Science and the Social Order* (Toronto: Collier Books, 1952), p. 137.

57. Bruno Latour, *Science in Action* (Cambridge, Mass.: Harvard University Press, 1987).

58. See, for example, Philip M. Boffey, "Scientists and Bureaucrats: A Clash of Cultures on FDA Advisory Panel," *Science* 191 (1976), pp. 144–146.

59. Adeline Levine, *Love Canal: Science, Politics, and People* (Lexington, Mass.: Lexington Books, 1982).

60. This issue was raised in connection with FDA's scientific review of the artificial sweetener aspartame. See Vincent Brannigan, "The First FDA Public Board of Inquiry: The Aspartame Case," in J. D. Nyhart and Milton M. Carrow, eds., *Law and Science in Collaboration* (Lexington, Mass.: Lexington Books, 1983), pp. 185–187.

61. Ronald Brickman, Sheila Jasanoff, and Thomas Ilgen, *Controlling Chemicals: The Politics of Regulation in Europe and the United States* (Ithaca, N.Y.: Cornell University Press, 1985).

62. For example, see the case study of formaldehyde regulation in the United States in Sheila Jasanoff, *Risk Management and Political Culture* (New York: Russell Sage Foundation, 1986), pp. 45–46.

63. Nicholas A. Ashford, "Advisory Committees in OSHA and EPA: Their Use in Regulatory Decisionmaking," *Science, Technology, and Human Values* 9 (1984), pp. 72–82.

64. Lock, *A Difficult Balance*, pp. 59–60.

65. See, for example, R. Jeffrey Smith, "Creative Penmanship in Animal Testing Prompts FDA Controls," *Science* 198 (1977), pp. 1227–1229. See also Broad and Wade, *Betrayers*, pp. 81–83.

5. EPA and the Science Advisory Board

1. 42 U.S.C. §4365–66.

2. Clean Air Act, §109(d).

3. This table is based on the SAB charter dated October 25, 1985. The charter was previously issued on October 31, 1981, and November 3, 1983.

4. House Committee on Science and Technology, *The Environmental Protection Agency's Research Program with Primary Emphasis on the Community Health and Environmental Surveillance System (CHESS): An Investigative Report* (hereinafter cited as *CHESS Investigative Report*), 94th Congress, 2nd Session (1976), pp. 103–105, 109.

5. Although the SAB commended the CHESS program for its pioneering research, the reviewers found fault with the selection of study populations, the definitions of "respiratory disease," and the methods of recording pollution levels. The report also criticized the EPA monograph for omitting basic data and caveats about the inferences to be drawn from the data. SAB, "Review of the CHESS Program," A Report of a Review Panel of the Science Advisory Board—Executive Committee, March 14, 1975, reprinted in House Committee on Science and Technology and House Committee on Interstate and Foreign Commerce, *The Conduct of the EPA's "Community Health and Environmental Surveillance System" (CHESS)*, 94th Congress, 2nd Session (1976), pp. 316–324.

6. *CHESS Investigative Report*, p. 5.

7. Id., p. 3.

8. Id., pp. 3, 14–15, 19, 102.

9. Id., p. 80.

10. Id., p. 19.

11. NAS/NRC, *Decision Making in the Environmental Protection Agency*, vol. II (Washington, D.C.: National Academy Press, 1977).

12. Id., p. 48.

13. H.R. 6232, §6(b)(2), 97th Congress, 2nd Session, May 18, 1982.

14. Ronald Reagan, Veto Message of S.2577, October 22, 1982.

15. R. Jeffrey Smith, "President Vetoes EPA R&D Bill," *Science* 218 (1982), p. 663.

16. GAO, *Attrition of Scientists and Engineers at Seven Agencies* (Washington, D.C.: GAO, 1984), p. 5.

17. R. Jeffrey Smith, "Gorsuch Strikes Back at EPA Critics," *Science* 217 (1982), p. 233.

18. Jonathan Lash, Katherine Gillman, and David Sheridan, *A Season of Spoils* (New York: Pantheon, 1984), p. 37.

19. Eliot Marshall, "Hit List at EPA?" *Science* 219 (1983), p. 1303.

20. Interview with SAB member, November 13, 1986, and personal communication from Terry Yosie, SAB Director.

21. These statistics were obtained from the Annual Report on Federal Advisory Committees and were supplied to the author by Mary Beeje, EPA Office of Management and Operations.

22. EPA, Report of the Director of the Science Advisory Board for Fiscal Year 1986, Washington, D.C., October 1986, p. 2.

23. Id.

24. Source: EPA, Report of the Director of the Science Advisory Board for Fiscal Year 1987, Washington, D.C., December 1987, p. 10.

25. Interests dissatisfied with this state of affairs include the American Industrial Health Council (AIHC), an influential lobbying group for the chemical industry. AIHC supports a more formal and technocratic selection process, with nominations officially solicited from the National Academy of Sciences and the National Institutes of Health. Interview with Robert Barnard (AIHC counsel), Washington, D.C., October 17, 1986.

26. For a description of this process, see EPA, "Science Advisory Board: Announcement of Procedure and Request for Nomination of Members," *Federal Register* 49, August 21, 1984, pp. 33169–33170.

27. EPA, 1986 SAB Annual report, p. 5.

28. Interview with Terry Yosie, Washington, D.C., October 17, 1986.

29. The policy of following such an openly political approach to science advice has attracted strong advocates. See Nicholas A. Ashford, "Advisory Committees in OSHA and EPA: Their Use in Regulatory Decisionmaking," *Science, Technology, and Human Values* 9 (1984), p. 78.

30. Interview with Terry Yosie, Washington, D.C., October 17, 1986.

31. This perhaps explains why the NAS recommendation of a two-year term for the SAB chairman has not been taken seriously by the agency. Another recommendation—that the SAB chairman should function officially as Science Adviser to the EPA administrator—has not been formally implemented, although Norton Nelson seems *de facto* to have performed in this capacity for Administrator Lee Thomas.

32. EPA, Science Advisory Board, "Future Risk: Research Strategies for the 1990s," Washington, D.C., September 1988, p. 2.

33. Id., p. 3.

34. Indeed, OTA has supported a shift from "end-of-pipe" controls to pollution prevention and reduction in the area of waste management. OTA, *Serious Reduction of Hazardous Waste* (Washington, D.C.: OTA, 1986).

35. Bruce N. Ames, "What Are the Major Carcinogens in the Etiology of Human Cancer? Environmental Pollution, Natural Carcinogens, and the Causes of Human Cancer: Six Errors," in Vincent T. De Vita, Jr., S. Hellman, and S. A. Rosenberg, eds., *Important Advances in Oncology, 1989* (Philadelphia: Lippincott, 1989), pp. 237–247.

36. Safe Drinking Water Act §1412(e).

37. Letter of Norton Nelson to Congressman John Dingell (presenting SAB comments on amendments to the Safe Drinking Water Act as they pertain to the additional scientific review responsibilities of the SAB), Washington, D.C., November 4, 1985.

38. Interview with SAB member, November 13, 1986.

6. The Science and Policy of Clean Air

1. The table is based on a report of the U.S. General Accounting Office: *Status of EPA's Air Quality Standards for Carbon Monoxide* (Washington, D.C.: GAO, 1984), pp. 2, 34–35.

2. For more detailed descriptions of the NAAQS standard-setting process, see Joseph Padgett and Harvey Richmond, “The Process of Establishing and Revising National Ambient Air Quality Standards,” *Journal of the Air Pollution Control Association* 33 (1983), pp. 13–16; Bruce C. Jordan, Harvey Richmond, and Thomas McCurdy, “The Use of Scientific Information in Setting Ambient Air Standards,” *Environmental Health Perspectives* 52 (1983), pp. 233–240.

3. Clean Air Act §108(a)(2).

4. Senate Committee on Environment and Public Works, *A Legislative History of the Clean Air Act Amendments of 1977,* 95th Congress, 2nd Session (August 1978), p. 6702.

5. Id., p. 6705.

6. See, for example, Jordan, Richmond, and McCurdy, “The Use of Scientific Information,” p. 235.

7. Id.

8. Appendix IV, EPA memorandum from Terry Yosie, Director, Science Advisory Board, to Frank Hanavan, Committee Management Officer, January 18, 1985.

9. R. Shep Melnick, *Regulation and the Courts: The Case of the Clean Air Act* (Washington, D.C.: Brookings Institution, 1983), pp. 281–294.

10. Id., p. 291.

11. *Federal Register* 42, December 30, 1977, p. 65264.

12. These members were Dr. Eileen G. Brennan (Rutgers), Dr. Edward F. Ferrand (New York City Department for Air Resources), Dr. Sheldon K. Friedlander (Cal Tech), Dr. Jimmye S. Hillman (University of Arizona), Dr. William W. Kellogg (National Center for Atmospheric Research, U.S. Department of Commerce), Dr. James N. Pitts, Jr. (University of California at Riverside), Dr. Robert Frank (University of Washington), Dr. James McCarroll (Stanford University Medical School), and Dr. Sheldon D. Murphy (University of Texas Medical School at Houston). In addition, Dr. Jay Jacobson was Liaison Representative from the U.S. Department of Energy.

13. API brief, *API v. Costle,* January 25, 1980, p. 20. See also Melnick, *Regulation and the Courts,* p. 286.

14. API brief, p. 29.

15. Lester B. Lave and Gilbert S. Omenn, *Cleaning the Air: Reforming the Clean Air Act* (Washington, D.C.: Brookings Institution, 1981), pp. 15–16.

16. Id., p. 11–12.

17. A. J. DeLucia and W. C. Adams, “Effects of O_3 Inhalation During Exercise on Pulmonary Function and Blood Biochemistry,” *Journal of Applied Physiology* 43 (1977), pp. 75–81.

18. Melnick, *Regulation and the Courts,* p. 286. See also EPA, “Revisions to the National Ambient Air Quality Standards for Photochemical Oxidants,” *Federal Register* 44, February 8, 1979, p. 8207.

19. *Federal Register* 44, p. 8207.

20. M. Granger Morgan of Carnegie-Mellon University notes that part of the problem with this technique was that EPA represented it as “decision analysis” even though, from the standpoint of professional decision analysts, it was neither a standard approach nor a technically sound one. Interview, Pittsburgh, March 15, 1988.

21. Melnick, *Regulation and the Courts,* p. 287.

22. API brief, p. 21.

23. According to one CASAC consultant, even in its earliest and least reliable version, EPA's method was much more than a "polling technique." Interview with M. Granger Morgan, Pittsburgh, March 15, 1988.

24. Note, however, that the D.C. Circuit held in 1987 in a case arising under §112 of the Clean Air Act that once EPA establishes a range of "safe" exposure levels, it may consider cost and technological feasibility in deciding precisely where to peg the emission standard for a hazardous air pollutant. *NRDC v. EPA,* 824 F.2d 1146 (D.C. Cir. 1986). This decision appeared to endorse the policy approach recommended by Schultze at the congressional hearings described below.

25. Senate Committee on Environment and Public Works, *Executive Branch Review of Environmental Regulations* (hereinafter cited as *Executive Branch Review*), 96th Congress, 1st Session (1979), p. 345.

26. See Melnick, *Regulation and the Courts,* p. 290.

27. *Executive Branch Review,* pp. 123–157 (NRDC comments on RARG/COWPS review of proposed standards).

28. In his testimony to Congress Kahn attempted to counter this viewpoint by arguing that there was no inherent conflict between health protection and cost considerations: "The popular conception that we must make choices between 'economic welfare' and environmental protection is simply wrong. Environmental values are economic values." *Executive Branch Review,* p. 338.

29. Pursuant to the Clean Air Act, NRDC should first have raised these objections before EPA by filing a petition for reconsideration. Since this procedure was not followed, the court held that it was not in a position to review the issues on their merits. *API v. Costle,* 665 F.2d 1176 (1981), pp. 1190–1192.

30. Id., p. 1190.

31. Id., p. 1189.

32. EPA, *Air Quality Criteria for Ozone and Other Photochemical Oxidants* (hereinafter cited as *Ozone Criteria Document*), Review Draft (1984), vol. II., p. 2-1.

33. Id., p. 2-5.

34. See *Ethyl Corp. v. EPA*., 541 F.2d 1 (D.C. Cir. 1976).

35. *Lead Industries Association, Inc. v. EPA,* 647 F.2d 1330 (D.C. Cir. 1980).

36. In adopting this position, the court showed an intuitive grasp of the problem of "experimenters' regress." As events proved, however, the mere likelihood of expert disputes was not a sufficient reason for withdrawing the issue of adverseness from consideration by experts.

37. CASAC Subcommittee on Ozone, Meeting Transcript (hereinafter cited as CASAC Transcript), March 5, 1985, pp. 158–159 (Crandall); March 6, 1985, pp. 18–19 (Koenig). Industry representatives present at the meeting strongly agreed with this position as well. See id., March 5, 1985, pp. 183, 211; March 4, 1985, p. 179.

38. CASAC Transcript, March 6, 1985, p. 32.

39. Id., pp. 9–10.

40. Id., p. 36.

41. Id., March 6, 1985, p. 4.

42. Id., p. 5.

43. Id.

44. *Inside EPA Weekly Report* 7, no. 17, April 25, 1986, p. 1.

45. Thomas McCurdy and Harvey Richmond, "Description of the OAQPS Risk Program and the Ongoing Lead NAAQS Risk Assessment Project" (paper presented at the 76th Annual Meeting of the Air Pollution Control Association, Atlanta, June 1983), p. 83-7.1. I am also indebted to M. Granger Morgan for providing a detailed account of these developments, in which he was an active participant.

46. See T. S. Wallsten and R. G. Whitfield, "Assessing the Risks to Young Children of Three Effects Associated with Elevated Blood Levels" (Argonne, Ill.: Argonne National Laboratory, December 1986).

47. CASAC Transcript, March 6, 1985, p. 7. It should be noted that Whittenberger's characterization of expert elicitation strikes those most familiar with the development of the technique after 1978 as either uninformed or unfair. M. Granger Morgan, for example, notes that the technique cannot in any sense be equated with a delphi approach. Interview, Pittsburgh, March 15, 1988.

48. CASAC Transcript, March 5, 1985, p. 173.

49. Id., March 4, 1985, p. 153.

50. Richmond and other proponents of the technique believe that it allows scientific experts to consider and weigh the full body of available evidence and to organize information in ways that permit balancing of incommensurables. See, for example, Harvey M. Richmond, "A Framework for Assessing Health Risks Associated with National Ambient Air Quality Standards," *Environmental Professional* 3 (1981), pp. 225–233. In the absence of techniques for formally assessing expert judgments, decisionmakers might be forced either to wait for nearly unattainable levels of objective scientific proof or to demand, under political pressure, that pollution sources prove the safety of their emissions as a precondition of operating.

51. CASAC Transcript, March 5, 1985, p. 34.

52. Id., March 6, 1985, p. 25.

53. Id., p. 27.

54. Id., pp. 23–24.

55. Id., p. 26.

56. These concerns may be even better justified when expert elicitation is used in connection with setting secondary ambient air quality standards, where the uncertainties are likely to be larger than in the health risk arena. Personal interview with ozone subcommittee member, Ithaca, N.Y., February 20, 1988.

57. Interview with CASAC member, Boston, February 18, 1987.

58. CASAC Transcript, March 4, 1985, p. 154.

59. *Ozone Criteria Document,* vol. V, ch. 13, pp. 13-70–13-72.

60. CASAC Transcript, March 4, 1985, pp. 179–180.

61. Id., March 5, 1985, p. 207.

62. Id., p. 206.

63. Id., pp. 147–148.

64. Id., pp. 151–152.

65. GAO, *EPA's Air Quality Standards for Carbon Monoxide*, pp. 6–7. The key study supporting the 1971 standard, by R. R. Beard and J. A. Wertheim, apparently demonstrated adverse effects on the central nervous system from low-level exposure to CO. Questions about the study were raised both before NAPCA in 1970 and before EPA in 1971. EPA concluded in 1979 that the study did not represent a sound scientific basis for standard-setting.

66. Id., p. 8. Letters questioning Aronow's results were published in *Annals of Internal Medicine,* where Aronow had published some of his work in 1972.

67. In one such action, EPA asked the Health Effects Institute (HEI), an independent research organization funded partly by EPA and partly by the automobile industry, to conduct a study of the effects of CO on angina patients. HEI's role in providing reliable data as a basis for regulatory decisionmaking is examined more fully in Chapter 11.

68. GAO, *EPA's Air Quality Standards for Carbon Monoxide,* p. 17. The CASAC closure letter indicated that EPA should treat the study by Einar Anderson with caution, but the committee found no reason to dispute the reported values.

7. Advisers as Adversaries

1. Rachel Carson, *Silent Spring* (New York: Houghton Mifflin, 1962).

2. Senate Committee on Agriculture and Forestry, Subcommittee on Agricultural Research and General Legislation, *Extension of the Federal Insecticide, Fungicide, and Rodenticide Act: Hearings on H.R. 8841,* 94th Congress, 1st Session (1975), p. 78 (statement of John C. Datt, American Farm Bureau Federation); p. 145 (statement of Paul S. Weller, National Council of Farmer Cooperatives); p. 173 (statement of John C. Stackhouse, National Association of State Departments of Agriculture); also pp. 7, 9, 10 (statements of J. Phil Campbell and Kenneth C. Walker, U.S. Department of Agriculture).

3. A Senate report noted that the purpose of creating the panel was "to assure that EPA obtains unbiased objective scientific opinion in making its decisions." Senate Committee on Agriculture and Forestry, *Extension of the Federal Insecticide, Fungicide, and Rodenticide Act,* S. Rep. No. 94–452, 94th Congress, 1st Session (1975), p. 9; see also House Committee on Agriculture, *Extension and Amendment of the Federal Insecticide, Fungicide, and Rodenticide Act,* H.R. Rep. No. 94–497, 94th Congress, 1st Session (1975), p. 11.

4. S. Rep. No. 94-452, p. 42 (statement of Peters D. Willson, National Wildlife Federation); p. 47 (statement of Linda M. Billings, Sierra Club); p. 55 (statement of Shirley A. Briggs and Samuel Epstein, The Rachel Carson Trust); p. 66 (statement of Stephanie Harris, Public Citizen Health Research Group).

5. See S. Rep. No. 94–452 at p. 14; House Committee on Agriculture, *Extension of the Federal Insecticide, Fungicide, and Rodenticide Act,* H.R. Rep. 94–668, 94th Congress, 1st Session (1975), pp. 4–5.

6. 7 U.S.C. §136w(d).

7. House Committee on Agriculture, *Federal Insecticide, Fungicide, and Rodenticide Act Extension,* H.R. Rep. 96-1020, 96th Congress, 2nd Session (1980), pp. 8, 17–18; House Committee on Agriculture, *Extension of Federal*

Insecticide, Fungicide, and Rodenticide Act: Hearings on H.R. 7018, 96th Congress, 2d Session (1980), p. 181 (statement of Hugh J. Wessinger, GAO).

8. 7 U.S.C. §136w(d). SAP thus has a right to comment on the agency's findings with respect to impacts on health and the environment, although EPA does not need the panel's concurrence in order to issue a suspension order.

9. 7 U.S.C. §136w(e).

10. 7 U.S.C. §136w(d).

11. Id.

12. Chevis Horne, unpublished paper prepared for Richard Merrill, University of Virginia Law School (1981).

13. Conference Report on S. 1678 H.R. Rep. 95-1560, 95th Congress, 2nd Session (1978), p. 43.

14. House Committee on Agriculture, *Federal Insecticide, Fungicide, and Rodenticide Act Extension,* H.R. Rep. 98-104, 98th Congress, 1st Session (1983), p. 9. Congress did not require EPA to appoint a new panel because the "already constituted panel was appointed pursuant to a charter under which all its activities will be conducted as if the requirements of section 24(d) and (e) govern the panel." Id.

15. Interview with Christopher Wilkinson, Ithaca, N.Y., December 11, 1985. See also John Walsh, "Spotlight on Pest Reflects on Pesticide," *Science* 215 (1982), p. 1595 (noting the Reagan administration's decision to dissolve the existing SAP and replace it with their own nominees).

16. The House Committee on Agriculture, for example, expressed concern that "the Scientific Advisory Panel may be functioning under adversary procedures attempting to act as an administrative law judge or as an arbitrator between scientists of conflicting opinions." H.R. Rep. 96–1020, pp. 4–5.

17. Horne, unpublished paper.

18. *Chemical Regulation Reporter* (BNA) 5, July 31, 1981, p. 399.

19. New pesticides may be registered only if the EPA administrator determines that they will not cause "unreasonable adverse effects on the environment" when used in accordance with accepted practice. FIFRA defines an "unreasonable adverse effect" to mean "any unreasonable risk to man or the environment, taking into account the economic, social, and environmental costs and benefits of the use of any pesticide." 7 U.S.C. §136(bb).

20. See, for example, House Committee on Government Operations, Environment, Energy and Natural Resources Subcommittee, *EPA's Pesticide Registration Activities* (hereinafter cited as *Pesticide Registration Hearings*), Part I, 98th Congress, 1st Session, September 26, 1983, p. 3 (testimony of Edwin L. Johnson, then director of OPP).

21. See, for example, Phillip L. Spector, "Regulation of Pesticides by the Environmental Protection Agency," *Ecology Law Quarterly* 5 (1976), pp. 244–252.

22. 7 U.S.C. §136d(b).

23. 7 U.S.C. §136w(d) provides that SAP should be asked to comment on regulatory proposals within the same time period as USDA.

24. 40 CFR 162.11 (1975).

25. See discussion of RPAR process in EPA, "Special Reviews of Pesticides;

Criteria and Procedures; Proposed Rule," *Federal Register* 50, March 27, 1985, p. 12192.

26. *Pesticide Registration Hearings,* Part I, pp. 31–32.

27. 7 U.S.C. §136a(c)(8).

28. *Federal Register* 50, pp. 12190–12191. A final rule governing special reviews was issued in November 1985.

29. For example, instead of the numerical criteria of acute toxicity that triggered RPAR review, EPA proposed a new criterion: "the use of a pesticide . . . may pose a risk of serious acute injury to humans or domestic animals." The rules also sought to clarify ambiguous language in the existing criteria, for example, the requirement that "multi-test evidence" will be used in evaluating mutagenic effects. *Federal Register* 50, pp. 12192, 12194.

30. House Committee on Government Operations, *Problems Plague the Environmental Protection Agency's Pesticide Registration Activities* (hereinafter cited as *Pesticide Report*), H.R. Rep. 98-1147, 98th Congress, 2nd Session, October 5, 1984, p. 18.

31. *Pesticide Report,* p. 19. Of course, by reregistering just one percent of the products, EPA could conceivably regulate a significantly higher percentage of uses and exposures.

32. Michael M. Simpson, "Ethylene Dibromide: History, Health Effects, and Policy Questions," Congressional Research Service, 84–622SPR, Washington, D.C., May 3, 1984, p. CRS-3.

33. Id., p. CRS-2.

34. David J. Hanson, "Agricultural Uses of Ethylene Dibromide Halted," *Chemical and Engineering News,* March 5, 1984, p. 13.

35. Simpson, "Ethylene Dibromide," p. CRS-3.

36. Philip M. Boffey, "Experts Are Split on Pesticide Risk," *New York Times,* February 4, 1984, p. 9.

37. Marjorie Sun, "EDB Contamination Kindles Federal Action," *Science* 223 (1984), p. 465.

38. Simpson, "Ethylene Dibromide," p. CRS-4.

39. California, Florida, Georgia, and Hawaii all found that groundwater in several areas was contaminated with EDB from soil fumigation. See Sun, "EDB Contamination," p. 464.

40. "Rebuttable Presumption against Registration and Continued Registration of Pesticide Products Containing Ethylene Dibromide (EDB)," *Federal Register* 42, December 14, 1977, pp. 63134-63148.

41. This proposal was set forth in a PD 2/3 published in the *Federal Register* on December 10, 1981.

42. SAP's comments were finally published in full as part of the emergency suspension order for EDB as a soil fumigant. See *Federal Register* 48, October 11, 1983, p. 46242.

43. *Pesticide Report,* p. 9.

44. Sun, "EDB Contamination," 466.

45. *Pesticide Report,* p. 6.

46. John Walsh, "Spotlight on Pest Reflects on Pesticide," pp. 1592–1596.

47. John Walsh, "EDB Causes a Regulatory Ripple Effect," *Science* 215 (1982), p. 1593.

48. Sun, "EDB Contamination," p. 464. See also Cass Peterson, "Patchwork of State Standards Complicates Federal EDB Effort," *Washington Post,* March 23, 1984, p. A2.

49. *Federal Register* 48, October 11, 1983, pp. 46228–46248.

50. *Federal Register* 49, February 6, 1984, pp. 4452–4457.

51. Statement by William D. Ruckelshaus, Administrator of the U.S. Environmental Protection Agency, on actions taken to control the pesticide EDB, Washington, D.C., February 3, 1984.

52. Statement by William D. Ruckelshaus, Administrator of the U.S. Environmental Protection Agency, on actions taken to control residues from the pesticide EDB on citrus fruit and papayas, Washington, D.C., March 2, 1984.

53. As one member of SAP observed at the meeting to review EPA's February 1984 suspension order, "The housewives, for instance . . . they feel that there was all of a sudden this great threat to the safety of the baked goods, cookies, and everything, and now that EDB has been suspended and canceled, everything is hunky-dory." Scientific Advisory Panel, Meeting Transcript (hereinafter cited as SAP Transcript), June 12, 1984, p. 77.

54. See, especially, SAP Transcript, November 29, 1983, pp. 42–49, 65–67.

55. Id., p. 42 (statement of Linda Vlier).

56. Id., p. 67 (statement of Christopher Wilkinson).

57. SAP Transcript, June 12, 1984, p. 49 (statement of Mr. Burin). The study referred to was conducted by Dutch government scientists L. H. J. C. Danse, F. L. van Felsen, and C. A. van der Heijden and appeared in *Toxicology and Applied Pharmacology,* February 1984.

58. SAP Transcript, June 12, 1984, p. 51 (statement of Robert Hollingworth).

59. Id., p. 63 (statement of Christopher Wilkinson).

60. See, for example, text at notes 73 and 77 below. See also SAP Transcript, June 12, 1984, p. 54 (statement of Stephen Sternberg).

61. William R. Havender, "EDB and the Marigold Option," *Regulation,* January/February 1984, p. 13.

62. SAP Transcript, November 29, 1983, p. 134 (statement of Christopher Wilkinson).

63. SAP Transcript, June 12, 1984, p. 71.

64. Id., p. 67.

65. Id., p. 52.

66. The record of the SAP meeting does not make it clear whether the panelists fully appreciated the statutory meaning of "emergency." Id., pp. 58–59 (statement of Paul Lapsley).

67. Id., p. 72.

68. For an authoritative statement of the problem, see NRC, *Risk Assessment in the Federal Government: Managing the Process* (Washington, D.C.: National Academy Press, 1983).

69. EPA, Ethylene Dibromide (EDB) Position Document 4, September 27, 1983, p. 72.

70. Hanson, "Agricultural Uses," p. 14.

71. Statements of Linda Randolph and Bailus Walker in House Committee on Government Operations, *Government Regulation of the Pesticide Ethylene Dibromide [EDB],* 98th Congress, 2nd Session (1984), pp. 176–177.

72. Id., pp. 77–78.

73. SAP Transcript, November 29, 1983, p. 120.

74. Id., p. 34.

75. Id., p. 67.

76. See, for example, EPA, Guidelines for Carcinogenic Risk Assessment, *Federal Register* 51, September 24, 1986, p. 33997 (approving dose-response assessment based on different tumor types and sites observed in same study).

77. SAP Transcript, November 30, 1983, p. 2.

78. EPA, Carcinogen Assessment Group, The Carcinogen Assessment Group's Evaluation of the Carcinogenicity of Dicofol (Kelthane), DDT, DDE, and DDD (TDE), EPA–600/6–85–002 (hereinafter cited as CAG Dicofol Evaluation), January 1985, p. 10.

79. *Pesticide Report,* p. 17.

80. House Committee on Government Operations, *Pesticide Registration Hearings,* Part 1, 98th Congress, 1st Session, September 26, 1983, and Part 2, 98th Congress, 2nd Session, June 7, 1984.

81. *Pesticide Report,* p. 17.

82. EPA, Proposed Notice of Intent to Cancel Dicofol Registrations, *Federal Register* 49 (1984), p. 39820.

83. CAG Dicofol Evaluation, pp. 2, 63.

84. Id., pp. 68–69.

85. Response of the Rohm and Haas Company to the Special Review for Products Containing Dicofol, EPA Docket No. OPP–30000.37A, February 15, 1985, pp. 13–14.

86. Id., p. 32.

87. Id., p. 42.

88. Id., p. 43.

89. Id., p. 42.

90. Id., pp. 2–3.

91. Id., p. 5.

92. Margaret Carlson, "Do You Dare to Eat a Peach?" *Nation* 133 (1989), p. 27.

93. Advertisement, *New York Times,* April 5, 1989, p. A11.

94. Bela Toth, "1,1-Dimethylhydrazine (Unsymmetrical) Carcinogens in Mice: Light Microscopic and Ultrastructural Studies on Neoplastic Blood Vessels," *Journal of the National Cancer Institute* 50 (1973), pp. 181–187.

95. The Toth studies of daminozide and UDMH were published, respectively, in *Cancer Research* 37 (1977), pp. 3497–3500 and *Cancer* 40 (1977), pp. 2427–2431. An inhalation study was carried out at the Air Force Aerospace Medical Research Laboratory, Wright Patterson Air Force Base, Ohio, in 1984.

96. See NRDC, "Chronology for Daminozide," in Daminozide Information Package, San Francisco, 1986 (hereinafter cited as Daminozide Chronology).

97. OPPE had to develop methods of measuring daminozide's intangible benefits (e.g., increased red color) and distinguishing between impacts on industry and on consumers. EPA, "The Economic Impacts and Social Benefits of Daminozide Use on Apples: An Empirical Study," (unpublished document, 1985).

98. Id., pp. 32–33. See also *Inside E.P.A. Weekly Report* 6, August 23, 1985, p. 8 (reporting on risk assessment as well as economic-impact assessment for daminozide).

99. Interview with EPA staff, Washington, D.C., October 22, 1986.

100. Interview with Steven Schatzow, Ithaca, N.Y., October 11, 1986; also see *Inside E.P.A.,* August 23, 1985, p. 8.

101. EPA, PD 2/3/4 on Daminozide, September 12, 1985.

102. SAP Transcript, September 26, 1985, p. 140.

103. Id., pp. 79–80, 95 (statement of Dr. Morris Cranmer, former director of National Center for Toxicological Research).

104. Id., p. 91. Another Uniroyal expert, Dr. Albert Kolbye, former Associate Director for Toxicological Sciences, Bureau of Foods, FDA, suggested that the effects seen in the Air Force Study may have been due to secondary toxic effects similar to those produced by irritants like formaldehyde. See id., pp. 104–105.

105. Id., pp. 88–90.

106. Id., p. 73 (statement of Dr. Stan D. Vesselinovitch) and p. 91 (statement of Morris Cranmer). See also comment of Gerald Schoenig: "but there is another common line where blood vessel tumors have been observed, and that is that the maximum tolerated dose has been exceeded in those studies." Id., p. 146.

107. Id., pp. 55, 59 (statement of Dr. Gerald Schoenig).

108. Id., pp. 47–48.

109. Id., p. 28 (statement of Dexter Goldman, EPA Office of Compliance Monitoring).

110. Id., p. 46.

111. Id., p. 47.

112. Id., p. 57.

113. Id., p. 73.

114. Id., p. 104.

115. Id., p. 100 (statement of Christopher Wilkinson).

116. Id., p. 158.

117. Id., p. 160.

118. Id., p. 97.

119. See "SAP Rejects EPA's Position on Alar; EPA Likely to Drop Cancellation for Now," Notes 5–12, *Pesticide and Toxic Chemical News,* October 2, 1985, p. 21.

120. Id., p. 28.

121. SAP, "Review of a Set of Scientific Issues Being Considered by EPA in Connection with the Special Review of Daminozide" (unpublished report, September 1985).

122. "OPP Officials Dismayed by FIFRA Scientific Advisory Panel's Alar Position," *Pesticide and Toxic Chemical News,* October 2, 1985, p. 28.

123. NRDC, Daminozide Chronology.

124. Irving Molotsky, "A Ban on Treated Apples," *New York Times,* July 27, 1986, p. 16.

125. NRDC et al., In re: Pesticide Tolerances for Daminozide—Petition for Establishment of Zero Tolerances, Washington, D.C., July 1, 1986.

126. *Ralph Nader v. EPA,* 859 F.2d 747 (9th Cir. 1988).

127. EPA, *Environmental News,* Washington, D.C., February 1, 1989. The final daminozide results confirmed interim reports submitted to EPA in 1988. At that time, the agency believed it did not have sufficient information to assess Alar's risks or to take steps toward cancellation. Barbara Rosewicz, "Chemicals Used on Fruit Tied in Test to Cancer," *Wall Street Journal,* April 21, 1988, p. 12.

128. Letter of John A. Moore to International Apple Institute, Washington, D.C., February 1, 1989.

129. John B. Oakes, "A Silent Spring, for Kids," *New York Times,* March 30, 1989, p. A25. Oakes, an NRDC trustee, was at one time editorial page editor of the *Times.*

130. James Warren, "Did Media Peel Away Facts in Apple Scare?" *Chicago Tribune,* March 26, 1989, p. 1.

131. Eliot Marshall, "Science Advisers Need Advice," *Science* 245 (1989), p. 20.

132. For a more extensive development of this argument, see Sheila Jasanoff, "EPA's Regulation of Daminozide: Unscrambling the Messages of Risk," *Science, Technology, and Human Values* 12 (1987), pp. 116–124.

133. One expert who was involved in NRDC's review process notes that the peer reviewers raised serious questions which the public interest group failed to address. Interview with John Wargo, Yale School of Forestry and Environmental Studies, New Haven, Conn., October 9, 1989. For additional examples of the political construction of peer review, see Sheila Jasanoff, "Contested Boundaries in Policy-Relevant Science," *Social Studies of Science* 17 (1987), pp. 195–230.

134. Steven Schatzow, "The Regulator's Dilemma—No Smoking Gun," *Natural Resources and Environment* 2 (1986), pp. 28–31, 62–66.

135. See, for example, Peter Huber, "Safety and the Second Best: The Hazards of Public Risk Management in the Courts," *Columbia Law Review* 85 (1985), pp. 277–337.

8. FDA's Advisory Network

1. Paul J. Quirk, "Food and Drug Administration," in James Q. Wilson, *The Politics of Regulation* (New York: Basic Books, 1980), pp. 196–199.

2. 21 U.S.C. §348(c)(3)(A) for food additives; §360b(d)(1)(H) for new animal drugs; §376(b)(5)(B)(ii) for color additives.

3. The structural reasons for these problems have been identified in an unending series of internal and external reports: inadequate remuneration, lack of professional opportunities and advancement, poor facilities, low morale. Peter B. Hutt, "Investigations and Reports Respecting FDA Regulation of New Drugs (Part I)," *Clinical Pharmacology and Therapeutics* 33 (1983), pp. 537–548.

4. For a listing of these committees, see 21 C.F.R. 14.100 (1986).

5. Interview with Joan Standaert, Washington, D.C., November 19, 1985.

6. No matter where the authority to make appointments is located, the possibility of making politically embarrassing selections can never be completely eliminated. In 1987, for instance, an NIH investigative team concluded that Dr. Jeffrey S. Borer, chairman of FDA's Cardio-Renal Committee, had made "unsupported statements" in a scientific publication. This concluded a six-year investigation into Borer's conduct at Cornell University Medical College. See David L.

Wheeler, "Government Increases Reliance on Universities to Detect and Probe Fraud by Own Researchers; Critics Are Wary," *The Chronicle of Higher Education* 34 (1987), p. A2. FDA apparently was unaware of the charges against Borer when it made the appointment.

7. House Committee on Government Operations, *Use of Advisory Committees by the Food and Drug Administration,* 93rd Congress, 2nd Session (1974) (hereinafter cited as *Fountain Hearings*).

8. See, for example, William M. Wardell, M. Hassar, S. N. Anavekar, and Louis Lasagna, "The Rate of Development of New Drugs in the United States, 1963 through 1975," *Clinical Pharmacology and Therapeutics* 24 (1978), pp. 133–45; William M. Wardell, "The Drug Lag Revisited: Comparison by Therapeutic Areas of Patterns of Drugs Marketed in the United States and Great Britain from 1972 through 1976," *Clinical Pharmacology and Therapeutics* 24 (1978), pp. 499–527.

9. *Fountain Hearings,* pp. 13–14.

10. 21 U.S.C. §355(e).

11. *Fountain Hearings,* p. 164 (statement by Dr. Julia Apter).

12. Id., p. 239.

13. Id., p. 132. The case is thus comparable to the Alar review discussed in the preceding chapter. When the SAP looked at the data on Alar case by case, it was persuaded that all of the studies were too flawed to be reliable.

14. Id., pp. 129, 163–164.

15. Id., p. 133.

16. In a curious wrinkle, this expert, Dr. Julia Apter, was also involved at about the same time in hearings before a Senate subcommittee, where she alleged conflicts of interest among members of an NIH committee. The NIH committee was in the process of reviewing a grant application for which Apter would have been the program director. When her application was denied, Apter brought suit claiming the denial was based, in part, on her testimony. See Steven Goldberg, "The Reluctant Embrace: Law and Science in America," *Georgetown Law Journal* 75 (1987), pp. 1356–1357.

17. *Fountain Hearings.*, p. 132.

18. Id., p. 225.

19. Id., p. 228. Crout's position, however, was strongly contradicted by Dr. Delphis Goldberg, a member of the subcommittee's professional staff.

20. Id., p. 131.

21. House Committee on Science and Technology, *The Food and Drug Administration's Process for Approving New Drugs,* 96th Congress, 1st Session (1979) (hereinafter cited as *New Drugs Hearings*), p. 530.

22. House Committee on Government Operations, *FDA's Regulation of Zomax,* H.R. Rep. 98-584, 98th Congress, 1st Session (1983), p. 2.

23. Id., p. 3.

24. Id., pp. 22–23.

25. Id., p. 25. Eight subcommittee members, however, dissented from the report's finding that FDA improperly referred the Zomax case to the advisory committee, and suggested that a fuller hearing should have been held to investigate this question. Id., pp. 31–32.

26. Seymour Shubin, "Triazure and Public Drug Policies," *Perspectives in*

Biology and Medicine (Winter 1979), pp. 185–204, reprinted in *New Drugs Hearings,* pp. 815–834.

27. Id., p. 821.

28. Id., p. 834.

29. Wardell, "Drug Lag Revisited," in *New Drugs Hearings,* p. 506.

30. Id., p. 508.

31. Interview with FDA official, Washington, D.C., November 19, 1985.

32. Telephone interview with Dr. Robert Temple, Director, Office of Drug Research and Review, Center for Drugs and Biologics, FDA, Washington, D.C., November 6, 1985. See also transcript of Cardio-Vascular and Renal Drugs Advisory Committee meeting, July 26, 1985, p. II-70 (comments by Drs. Harrison and Borer).

33. Id., p. II-95.

34. See, for example, id., pp. II-99–II-100 (comments by Drs. Lavy and Borer).

35. See, for example, id., pp. II-106–II-110.

36. Id., p. II-66.

37. Id., pp. II-85, II-110.

38. For a discussion of European approaches to using advisory committees, see Ronald Brickman, Sheila Jasanoff, and Thomas Ilgen, *Controlling Chemicals: The Politics of Regulation in Europe and the United States* (Ithaca, N.Y.: Cornell University Press, 1985).

39. Wardell, "Drug Lag Revisited," p. 508.

40. The Cardiac Arrhythmia Suppression Trial (CAST) Investigators, "Preliminary Report: Effect of Encainide and Flecainide on Mortality in a Randomized Trial of Arrhythmia Patients," *New England Journal of Medicine* 321 (1989), pp. 406–412.

41. See, for example, correspondence on this issue in *New England Journal of Medicine* 321 (1989), pp. 1754–1756.

42. Cardio-Vascular and Renal Drugs Advisory Committee, Transcript of Proceedings, Bethesda, Md., October 5, 1989, p. 183.

43. Id., p. 170.

44. 21 C.F.R. 14.100(a); also see Alan R. Bennett, "Committee or Commissioner," *Food Drug Cosmetic Law Journal* 32 (1977), p. 323.

45. FASEB, Life Sciences Research Office (LSRO), *The Reexamination of the GRAS Statutes of Sulfiting Agents* (Washington, D.C.: FASEB, 1985) (hereinafter cited as *FASEB Sulfites Report*), p. 5.

46. 21 U.S.C. §321(s).

47. FASEB, Select Committee on GRAS Substances, "Evaluation of Health Aspects of GRAS Food Ingredients: Lessons Learned and Questions Unanswered," *Federation Proceedings* 36, no. 11 (1977), p. 2528 (hereinafter cited as *SCOGS Report*).

48. *FASEB Sulfites Report,* p. 1.

49. Oral statement of Michael Jacobson, Executive Director, CSPI, before House Committee on Energy and Commerce, March 27, 1985.

50. *FASEB Sulfites Report,* p. 2.

51. Id., p. iii.

52. Transcript of Open Meeting, FASEB Ad Hoc Review Panel on Sulfiting Agents, Bethesda, Md., November 29, 1984, p. 9 (hereinafter cited as Transcript). See also "FDA Advisory Panel Refuses to Recommend Curbs on Sulfiting Agents," *Nutrition Action,* December 1984, p. 3.

53. Transcript, p. 12.

54. Id., p. 82. See also *SCOGS Report,* p. 2547.

55. Transcript, p. 79.

56. Id., p. 30.

57. *FASEB Sulfites Report,* p. 59.

58. Data presented by a researcher from the University of Wisconsin suggested in fact that 1–2 percent of all asthmatics might be sensitive to sulfites. Id., p. 49.

59. Id., p. 60.

60. *Federal Register* 51, July 9, 1986, pp. 25021–26876.

61. Id., pp. 25012–25020.

62. Interview with Mitchell Zeller, CSPI, Washington, D.C., December 13, 1985.

63. 21 U.S.C. §376.

64. 21 U.S.C. §376(b)(5)(B).

65. House Committee on Government Operations, *The Regulation by the Department of Health and Human Services of Carcinogenic Color Additives,* 98th Congress, 2nd Session (1984) (hereinafter cited as *Color Additives Hearings*), pp. 109–114 (statement of Commissioner Young).

66. Id., p. 111.

67. 613 F2d. 947 (D.C. Cir. 1979).

68. See *Color Additives Hearings,* p. 28, for CTFA listing of FDA decisions establishing *de minimis* risk policy.

69. Id., pp. 237–271.

70. Id., p. 248.

71. *Color Additive Hearings,* pp. 314–327, 358–360.

72. Id., pp. 411–412.

73. Id., p. 412.

74. Ronald W. Hart et al., "Final Report of the Color Additive Scientific Review Panel," *Risk Analysis* 6, no. 2 (1986) (hereinafter cited as "FDA Color Report"), p. 118. The members of the panel were Ronald Hart (NCTR), Stan Freni (DEHHE, CDC), David Gaylor (NCTR), James Gillette (National Heart, Lung and Blood Institute), Larry Lowry (NIOSH), Jerrold Ward (NCI), Elizabeth Weisburger (NCI), Paul Lepore (FDA), and Angelo Turturro (NCTR).

75. "FDA Color Report," p. 118.

76. Panel members included scientists from NCTR, CDC, NIOSH, NCI, and the National Heart, Lung and Blood Institute.

77. "FDA Color Report," p. 118.

78. "New Panel to Review Safety of 6 Dyes," *New York Times,* May 29, 1985, p. C7.

79. 831 F.2d 1108 (D.C. Cir. 1987).

80. Id., p. 1112.

81. Id., p. 1121.

9. Coping with New Knowledge

1. *Federal Register* 51, September 24, 1986, pp. 33992–34003.

2. For an extended account of these developments, see Nathan J. Karch, "Explicit Criteria and Principles for Identifying Carcinogens: A Focus of Controversy at the Environmental Protection Agency," in NAS-NRC, *Decision Making in the Environmental Protection Agency,* vol. 2a (Washington, D.C.: National Academy Press, 1977), pp. 119–206.

3. Id., p. 138.

4. *Federal Register* 41, May 25, 1976, pp. 21402–21405.

5. *Federal Register* 44, October 10, 1979, pp. 58642–58661.

6. *Journal of the National Cancer Institute* 63 (1979), pp. 241–268.

7. *Federal Register* 43, June 13, 1978, pp. 25658–25665.

8. *Federal Register* 45, January 22, 1980, pp. 5002–5296.

9. *Dow Chemical v. CPSC,* 459 F.Supp. 378 (W.D. La. 1978).

10. OSHA's inability to find scientifically justifiable formulations consistent with its administrative objectives is discussed in Sheila Jasanoff, "Science and the Limits of Administrative Rulemaking: Lessons from the OSHA Cancer Policy," *Osgoode Hall Law Journal* 20 (1982), pp. 536–561.

11. Id., pp. 554–555.

12. Terry Yosie, "The Culture of Risk Assessment at the Environmental Protection Agency," paper presented at the Annual Meeting of the American Chemical Society, New York, April 15, 1986, p. 8.

13. Terry Yosie, "EPA's Risk Assessment Culture," *Environmental Science and Technology* 21 (1987), p. 529.

14. AIHC, "AIHC Proposal for a Science Panel" (mimeo, Scarsdale, N.Y., March 1981).

15. NAS-NRC, *Risk Assessment in the Federal Government: Managing the Process* (Washington, D.C.: National Academy Press, 1983).

16. Id., pp. 166–167, 171–175. This Board was to be a congressionally chartered entity independent of the regulatory agencies. The panel noted that OSTP did not meet these criteria because it was located in the Executive Office of the President and hence likely to lack independence.

17. OSTP formed an Interagency Staff Group on Carcinogens specifically to address Recommendations 5 through 8 of the 1983 NAS risk-assessment report.

18. Interview with SAB member, New York, February 4, 1987.

19. See Eliot Marshall, "EPA's High-Risk Carcinogen Policy," *Science* 218 (1982), pp. 975–978; Allen Morrison, "Risk Analysis: Is EPA Changing the Rules," *Civil Engineering—ASCE* (October 1982), pp. 60–63.

20. House Committee on Energy and Commerce, Subcommittee on Commerce, Transportation and Tourism, *Control of Carcinogens in the Environment,* 98th Congress, 1st Session (March 1983), p. 72.

21. Ronald W. Hart and Angelo Turturro, "History, Status and Potential Impact of the White House Office of Science and Technology Policy Cancer Guidelines," paper presented at New Orleans meeting of Mid-South Chapter of the American Association of Industrial Hygienists, October 1984.

22. Source: id.

23. OSTP, "Chemical Carcinogens; Notice of Review of the Science and Its Associated Principles," *Federal Register* 49, May 22, 1984, pp. 21594–21661.

24. Interview with SAB member, New York, February 4, 1987.

25. *Federal Register* 50, March 14, 1985, pp. 10371–10442.

26. *Environmental Health Perspectives* 67 (1986), pp. 201–282.

27. *Federal Register* 50, p. 10376.

28. Terry Yosie, Executive Director of SAB, notes that within the past five years EPA has developed many centers of risk assessment, so that CAG's influence has decreased. See Terry Yosie, "The Culture of Risk Assessment at the Environmental Protection Agency," paper presented at Annual Meeting of the American Chemical Society, April 15, 1986, pp. 7–8.

29. "Report by the SAB Carcinogenicity Guidelines Review Group," transmitted with letter from Norton Nelson, Chairman, SAB, to Lee Thomas, June 19, 1985.

30. The SAB reviewers felt that EPA should more fully address (1) transplacental multigenerational or total lifetime bioassays and (2) sensitivity analysis of quantitative estimates of potency.

31. See National Academy of Public Administration, *Presidential Management of Rulemaking in Regulatory Agencies* (Washington, D.C., January 1987).

32. In 1986 OIRA exceeded the 90-day limit 69 times. *OMB Watch,* January 28, 1987, p. 1.

33. *EDF. v. Thomas,* 627 F.Supp. 566 (D.D.C. 1986).

34. U.S. House of Representatives, Committee on Energy and Commerce, Subcommittee on Oversight and Investigations, *OMB Review of FDA Regulations,* 99th Congress, 2nd Session (1986), p. 1.

35. Id., p. 15.

36. U.S. House of Representatives, Committee on Government Operations, Subcommittee on Intergovernmental Relations and Human Resources, *The Regulation by the Department of Health and Human Services of Carcinogenic Color Additives,* 98th Congress, 2nd Session (1984), pp. 437–450.

37. *Public Citizen Health Research Group v. Tyson,* 796 F.2d 1479 (D.C. Cir. 1986).

38. *OMB Watch,* August 25, 1986.

39. *OMB Watch,* February 26 and November 26, 1986.

40. U.S. House of Representatives, Committee on Energy and Commerce, *OMB Review of CDC Research: Impact of the Paperwork Reduction Act,* Committee Print 99-MM, 99th Congress, 2nd Session (1986).

41. The scientist, Dr. James E. Hansen, emphasized the illegitimacy of such intervention through boundary work: "I should be allowed to say what is my scientific position; there is no rationale by which O.M.B. should be censoring scientific opinion. I can understand changing policy, but not science." Philip Shabecoff, "Scientist Says Budget Office Altered His Testimony," *New York Times,* May 8, 1989, p. A1.

42. U.S. House of Representatives, Committee on Science and Technology, *H.R. 4192: The Risk Assessment Research and Demonstration Act of 1983,* 98th Congress, 2nd Session (1984), pp. 289–437.

43. Eliot Marshall, "OMB and Congress at Odds over Cancer Risk Policy," *Science,* 233 (1986), p. 618.

44. Wendy L. Gramm, Letter to Lee Thomas, August 12, 1986.

45. Lee M. Thomas, Letter to Wendy L. Gramm, August 22, 1986.

46. See, in particular, John D. Graham, Laura C. Green, and Marc J. Roberts, *In Search of Safety: Chemicals and Cancer Risk* (Cambridge, Mass.: Harvard University Press, 1988).

47. The CIIT study exposed the test animals to three dose levels. No carcinomas were observed at 2 ppm, 2 at 5.6 ppm, and 103 at 14.5 ppm. W. D. Kerns, K. L. Pavkov, D. J. Donofrio, E. J. Gralla, and J. A. Swenberg, "Carcinogenicity of Formaldehyde in Rats and Mice after Long-term Inhalation Exposure," *Cancer Research* 43 (1983), pp. 4382–4392.

48. "Report of the Federal Panel on Formaldehyde," *Environmental Health Perspectives,* 43 (1982), pp. 139–168.

49. Id., p. 165.

50. Sheila Jasanoff, "Cultural Aspects of Risk Assessment in Britain and the United States," in Branden B. Johnson and Vincent T. Covello, eds., *The Social and Cultural Construction of Risk* (Dordrecht, Netherlands: Reidel, 1987), pp. 359–397.

51. John D. Graham, "Federal Regulation of Formaldehyde: The Agencies, Their Statutes, and Recent Decisions" (unpublished paper, Scientific Conflict Mapping Project, Harvard School of Public Health, March 1984).

52. *Federal Register* 47, April 2, 1982, pp. 14366–14421.

53. *Gulf South Insulation v. CPSC,* 701 F.2d 1137 (1983).

54. Id., p. 1146.

55. See, for example, Kenneth S. Abraham and Richard A. Merrill, "Scientific Uncertainty in the Courts," *Issues in Science and Technology* 2 (1986), pp. 93–107.

56. Devra L. Davis, "The 'Shotgun Wedding' of Science and Law: Risk Assessment and Judicial Review," *Columbia Journal of Environmental Law* 10 (1985), pp. 67–109.

57. Bette Hileman, "Formaldehyde," *Environmental Science and Technology* 16 (1982), p. 544A.

58. Nicholas A. Ashford, C. William Ryan, and Charles C. Caldart, "Law and Science Policy in Federal Regulation of Formaldehyde," *Science* 222 (1983), p. 896.

59. For an overview of this debate, see House Committee on Science and Technology, *Formaldehyde: Review of Scientific Basis of EPA's Carcinogenic Risk Assessment,* 97th Congress, 2nd Session (1982).

60. Graham et al., *In Search of Safety,* p. 213.

61. Deliberations of the Consensus Workshop on Formaldehyde, Little Rock, Ark., October 3–6, 1983, p. 47.

62. Id., p. 139.

63. "Science" can be understood in this context either as the scientific community or as scientific knowledge. From the point of view of federal regulators, either a consensus among scientists or the "smoking gun" of definitive evidence of harm would provide extremely desirable relief from controversy.

64. "U.S. Study Disputes Cancer Risk of Formaldehyde," *New York Times,* March 3, 1986, p. A1; Peter Perl, "Cancer Link Conclusions in Dispute," *Washington Post,* March 3, 1986, p. A1.

65. House Committee on Energy and Commerce, Subcommittee on Oversight and Investigations, *Formaldehyde Study,* 99th Congress, 2nd Session (1986), pp. 48–49.

66. Id., p. 116.

67. Id., p. 228 (statement of Margaret Seminario, AFL-CIO).

68. Blair insisted, for example, that NCI had retained "complete control" over the study, and Dr. Maureen O'Berg, Dupont's director of epidemiology and a coauthor of the study, indicated that the involvement of the corporate authors was largely limited to review and editing. Perl, "Cancer Link Conclusions," p. A8.

69. Graham et al., *In Search of Safety,* p. 72.

70. Richard A. Griesemer, Chair, SAB Environmental Health Committee, and Norton Nelson, Chair, SAB Executive Committee, Letter to Lee M. Thomas, EPA Administrator, October 1, 1985, p. 2.

71. Interview with Joseph Merenda, Director, Existing Chemicals Division, EPA Office of Toxic Substances, November 20, 1985. There apparently were some disagreements within the agency about the significance of SAB's criticisms, with OTS taking the position that SAB's comments indicated basic support for the agency. Interview with David Dull, Deputy Director, Chemical Control Division, November 20, 1985.

72. Response by the Environmental Health Committee of EPA's Science Advisory Board to a List of Questions Submitted by the Office of Pesticides and Toxic Substances Regarding a Draft Document, "Preliminary Assessment of Health Risks to Garment Workers and Certain Home Residents from Exposures to Formaldehyde," October 1, 1985.

73. Dale Hattis, "The Use of Biological Markers in Risk Assessment" (unpublished paper, MIT, Cambridge, Mass., 1986).

74. Mercedes Casanova-Schmitz, Thomas B. Starr, and Henry D'A. Heck, "Differentiation between Metabolic Incorporation and Covalent Binding on the Labeling of Macromolecules in the Rat Nasal Mucosa and Bone Marrow Inhaled (14C)- and (3H)CH_2O," *Toxicology and Applied Pharmacology* 76 (1984), pp. 26–44.

75. "Expert Panel Report on HCHO Pharmacokinetic Data and CIIT Response," Appendix I in EPA Office of Toxic Substances, "Assessment of Health Risks to Garment Workers and Certain Home Residents from Exposure to Formaldehyde," April 1987.

76. Letter of Robert A. Neal, President, CIIT, to William H. Farland, Deputy Director, OPTS Health and Environmental Review Division, February 4, 1986 (enclosing "Comments on the Final Report of the Panel Reviewing the CIIT Pharmacokinetic Data on Formaldehyde" by M. Casanova, T. B. Starr, and H. D'A. Heck).

77. EPA Office of Pesticides and Toxic Substances, "Assessment of Health Risks to Garment Workers and Certain Home Residents from Exposure to Formaldehyde" (hereinafter cited as Formaldehyde Risk Assessment), Washington, D.C., April 1987.

78. Formaldehyde Risk Assessment, pp. 4-67–4-68.

79. The agency was, of course, also assisted by congressional support, in particular, Representative Dingell's threat to cut off funding for OIRA. This support, however, might not have been as easily forthcoming if EPA's scientific

advisers had not already indicated their support for the risk-assessment guidelines.

80. See, for instance, "Formaldehyde Textile Issues Referral to OSHA This Week," *Pesticide and Toxic Chemical News,* March 19, 1986, p. 20.

81. Sheila Jasanoff, *Risk Management and Political Culture* (New York: Russell Sage Foundation, 1986), pp. 41–53.

10. Technocracy Revisited

1. Eliot Marshall, "Health Effects Institute Links Adversaries," *Science* 227 (1985), pp. 729–730.

2. Id., p. 729.

3. General Accounting Office (GAO), *Air Quality Standards—The Role of the Health Effects Institute in Conducting Research* (Washington, D.C.: GAO, 1986), p. 17.

4. Source: GAO, *The Role of the HEI,* p. 9.

5. HEI, *The Health Effects Institute Annual Report* (Cambridge, Mass.: HEI, 1985), p. 6.

6. Id., p. 8.

7. Id., p. 13.

8. Id., p. 14.

9. Personal communication, Rashid Shaikh, HEI, Cambridge, Mass., March 23, 1988.

10. Interview with Andrew Sivak, Cambridge, Mass., June 6, 1988.

11. Id., p. 19.

12. Id., p. 22.

13. Report of the Institutes's Health Review Committee in HEI, *Acute Effects of Carbon Monoxide Exposure on Individuals with Coronary Artery Disease,* Research Report no. 25 (Cambridge, Mass.: 1989), pp. 81–98. See also Elizabeth N. Allred et al., "Short-term Effects of Carbon Monoxide Exposure on the Exercise Performance of Subjects with Coronary Artery Disease," *New England Journal of Medicine* 321 (1989), pp. 1426–1432.

14. GAO, *The Role of the HEI,* pp. 1, 21.

15. Id., p. 32.

16. Interviews with HEI staff, Cambridge, Mass., December 1986.

17. GAO, *Air Quality Standards—EPA's Standard Setting Process Should Be More Timely and Better Planned* (Washington, D.C.: GAO, 1986), p. 4.

18. HEI Asbestos Study Team Final Report, Cambridge, Mass., March 1989.

19. *HEI Insight* 5 (1989), p. 1.

20. Warren E. Whyte, "Effectiveness of the NAS-NRC Drug Effectiveness Review," *Food Drug Cosmetic Law Journal* 25 (1970), p. 92.

21. Rodney R. Munsey, "Private Remedies, Including Class Actions, Arising out of Implementation of NAS-NRC Drug Efficacy Review," *Food Drug Cosmetic Law Journal* 26 (1971), p. 669.

22. The classification categories were "effective," "probably effective," "possibly effective," "ineffective," "effective, but," and "ineffective as a fixed combination." Id.; also see Whyte, "Effectiveness," p. 92.

23. Whyte, "Effectiveness," pp. 92–93; see also Vincent A. Kleinfeld, "Some Reflections on Selected Aspects of NAS/NRC Review," *Food Drug Cosmetic Law Journal* 27 (1972), p. 140.

24. Munsey, "Private Remedies," pp. 670–671.

25. Id., p. 669.

26. Whyte, "Effectiveness," pp. 97–98.

27. Remarks of Dr. Daniel X. Freedman, quoted in Joseph D. Cooper, ed., *Decision-Making on the Efficacy and Safety of Drugs* (Washington, D.C.: Interdisciplinary Communications Associates, 1971), p. 108.

28. Id., pp. 93–96; also see Stanley L. Temko, "Litigation Resulting from NAS-NRC Review and FDA Demands for 'New Drug' Status," *Food Drug Cosmetic Law Journal* 26 (1971), pp. 464–471.

29. Temko, "Litigation," pp. 465–466.

30. Cooper, *Decisionmaking,* p. 107.

31. "Federation of American Societies for Experimental Biology," FASEB leaflet, Bethesda, Md.

32. "Life Sciences Research Office," FASEB leaflet, Bethesda, Md.

33. LSRO, *Insights on Food Safety Evaluation* (Bethesda, Md.: LSRO, 1982), p. 6.

34. SCOGS, "Evaluation of Health Aspects of GRAS Food Ingredients: Lessons Learned and Questions Unanswered," *Federation Proceedings* 36 (1977), p. 2530.

35. LSRO, *Insights,* p. 6.

36. Source: id., p. 5.

37. SCOGS, "Evaluation," p. 2529.

38. LSRO, *Insights,* p. 9.

39. SCOGS, "Evaluation," p. 2547.

40. LSRO, *Insights,* p. 27.

41. Id., p. 26.

42. Id., p. 33.

43. Todd R. Smyth, "The FDA's Public Board of Inquiry and the Aspartame Decision," *Indiana Law Journal* 59 (1983), pp. 628–630.

44. Vincent Brannigan, "The First FDA Public Board of Inquiry: The Aspartame Case," in J. D. Nyhart and Milton M. Carrow, eds., *Law and Science in Collaboration* (Lexington, Mass.: Lexington Books, 1983), p. 181.

45. *Community Nutrition Institute (CNI) v. Young,* 773 F.2d 1356 (D.C. Cir. 1985), pp. 1358–1359.

46. The members were Walle Nauta (psychology and brain science) and Vernon Young (nutritional biochemistry) of MIT, and Deter Lampert (pathology) of the University of California at San Diego. One critic has suggested that it was an error not to appoint a toxicologist to the PBOI and that the board would correctly have concluded that aspartame does not induce brain tumors if it had included a competent toxicologist.

47. Vincent Brannigan describes this result as "absurd" and argues that it would be better practice to ask for evaluations of the panel as constituted than to entertain objections concerning individual appointees. "The First FDA Public Board," p. 187.

48. *Federal Register* 46, July 24, 1981, p. 38286.

49. Id., p. 38284–38308.

50. 773 F.2d 1356, pp. 1364–1367.

51. *Food Drug Cosmetic Law Reports* (New Developments), August 18, 1986, p. 45207.

52. As of January 1, 1987, FDA had received 3,133 consumer complaints through its postmarket surveillance program on aspartame. *Food Chemical News,* January 5, 1987, p. 34. The agency subsequently referred the issue of allergic reactions to aspartame to its Ad Hoc Advisory Committee on Hypersensitivity to Food Constituents, *Federal Register* 51, April 11, 1986, pp. 12570–12571.

53. See *Food Chemical News* 29, June 8, 1987, p. 47.

54. Brannigan, "The First FDA Public Board," pp. 182–185. Smyth is less concerned about the delegation of responsibility for policy to the board, but apparently chiefly because in this case the FDA Commissioner showed that he was not bound by the board's judgment. "The FDA's Public Board," p. 639.

55. Examples include FDA's actions on drugs such as propranolol and triazure. See Chapter 8 for additional details.

56. Smyth, "The FDA's Public Board," p. 647. Brannigan argues that "the entire rationale for convening a board of experts collapses if their opinion carries no more weight than that of an administrative law judge." "The First FDA Public Board," p. 198.

57. Smyth, "The FDA's Public Board," p. 648.

58. Nyhart and Carrow, *Law and Science,* p. 205.

11. The Political Function of Good Science

1. The National Research Council in effect repudiated this idea in its influential study of federal risk-assessment practices. NRC, *Risk Assessment in the Federal Government: Managing the Process* (Washington, D.C.: National Academy Press, 1983). Other works that cast doubt on the feasibility of separating science from policy in the regulatory process include Liora Salter, *Mandated Science* (Dordrecht, Netherlands: Kluwer, 1988), and Mark E. Rushefsky, *Making Cancer Policy* (Albany: SUNY Press, 1986).

2. William Lowrance, *Of Acceptable Risk* (Los Altos, Calif.: William Kaufmann, 1976).

3. The metaphor of "ripeness" in administrative law describes the status of an agency decision that is definitive enough to be judicially reviewed. While such a state may be more difficult to identify for purposes of scientific review, it is clear that attempts to foreclose regulatory debates with the aid of expert advice will fail unless the technical issues are sufficiently defined and sharpened and relevant data are available.

4. As I have noted elsewhere, decisionmakers may gain as much from labeling an issue "policy" as scientists do from labeling the issue "science." Sheila Jasanoff, "Contested Boundaries in Policy-Relevant Science," *Social Studies of Science* 17 (1987), pp. 195–230.

5. See, for instance, the advertisement headed "Our Food Supply is Safe," *New York Times,* April 5, 1989, p. A11.

6. At best, the committee might identify certain features of such studies—for example, the lack of replication—as troublesome. But the committee will see this as a problem only if the record contains explicit indications that other researchers have tried and failed to replicate key experiments.

7. *Ethyl Corp. v. EPA,* 541 F.2d 1 (D.C. Cir. 1976).

8. EPA has been criticized for such overreliance on a limited range of scientific views on other occasions as well, most notably in its dealings with Umberto Saffiotti and Roy Albert on issues of cancer risk assessment (see Chapter 9).

9. As noted in Chapter 5, EPA has used its discretionary power to appoint SAB members to create an informal balance of interests on that key committee.

10. Thus, drug advisory committee meetings are attended chiefly by manufacturers and stock market analysts. Interview with Joan Standaert, FDA, November 19, 1985.

11. Such a procedure might have helped prevent the Alar debacle. This was the view of Christopher Wilkinson, a former SAP member. Interview, Ithaca, N.Y., December 11, 1985.

Index